W0107577

# Computer Integrated Manufacturing

## Volume I: Revolution in Progress

Robert U. Ayres

*International Institute for Applied Systems Analysis*
*Laxenburg, Austria*

Springer-Science+Business Media, B.V.

| | |
|---|---|
| UK | Chapman & Hall, 2–6 Boundary Row, London SE1 8HN |
| USA | Van Nostrand Reinhold, 115 5th Avenue, New York NY10003 |
| JAPAN | Chapman & Hall Japan, Thomson Publishing Japan, Hirakawacho Nemoto Building, 7F, 1–7–11 Hirakawa-cho, Chiyoda-ku, Tokyo 102 |
| AUSTRALIA | Chapman & Hall Australia, Thomas Nelson Australia, 102 Dodds Street, South Melbourne, Victoria 3205 |
| INDIA | Chapman & Hall India, R. Seshadri, 32 Second Main Road, CIT East, Madras 600 035 |

First edition 1991

© 1990 Springer Science+Business Media Dordrecht

Softcover reprint of the hardcover 1st edition 1990

Originally published by International Institute for Applied Systems Analysis CIM in 1990.

ISBN 978-94-015-1108-7       ISBN 978-94-015-1106-3 (eBook)
DOI 10.1007/978-94-015-1106-3

All rights reserved. No part of this publication may be reproduced or transmitted, in any form or by any means, electronic, mechanical, photocopying, recording or otherwise, or stored in any retrieval system of any nature, without the written permission of the copyright holder and the publisher, application for which shall be made to the publisher. The publisher makes no representation, express or implied with regard to the accuracy of the information contained in this book and cannot accept any legal responsibility or liability for any errors or omissions that may be made.

**British Library Cataloguing in Publication Data**
Ayres, R.U.
   Computer integrated manufacturing.
   Vol. 1: Revolution in progress
   1. Computer integrated manufacturing systems
   I. Title
   670.285


**Library of Congress Cataloging-in-Publication Data**
Ayres, Robert U.
   Computer integrated manufacturing/Robert U. Ayres.
      p.    cm.
   Includes bibliographical references and index.
   Contents: Contents: v. 1. Revolution in progress

   1. Computer integrated manufacturing systems. I. Title.
   TS155.6.A98 1991
   670.285—dc20
                                                        90-21365
                                                        CIP

Volume I of a projected four-volume set. Forthcoming volumes include:

Computer Integrated Manufacturing, Volume II: The Past, the Present, and the Future
Edited by *R.U. Ayres, W. Haywood, M.E. Merchant, J. Ranta, and H.-J. Warnecke*

Computer Integrated Manufacturing, Volume III: Models, Case Studies, and Forecasts of Diffusion
Edited by *R.U. Ayres, W. Haywood and Iouri Tchijov*

Computer Integrated Manufacturing, Volume IV: Economic and Social Impacts
Edited by *R.U. Ayres, R. Dobrinsky, W. Haywood, K. Uno, and E. Zuscovitch*

# Computer Integrated
# Manufacturing

# Contents

# Preface

## Motivation of the Study

CIM is an abbreviation for Computer Integrated Manufacturing (see Section 1.4, Chapter 1). It is an acronym that has become fairly well known in recent years in manufacturing and related engineering circles, although most laymen have never heard the term. The question naturally arises: why should a multinational, multidisciplinary organization like IIASA undertake a major study of such a seemingly arcane topic at this time of *glasnost*, *perestroika*, nuclear disarmament, and concern with global climate change, among other issues of global significance competing for our attention?

The short answer to this rhetorical question is that we think a new industrial revolution is under way. If the first industrial revolution was the substitution of steam power for human and animal muscles, the present revolution is the substitution of electronic sensors for human eyes and ears, and computers for human brains, at least in a certain category of routine 'on-line' manufacturing operations. Its consequences may be as far-reaching (and as unexpected) as the consequences of the 'first' industrial revolution, which was just beginning about two centuries ago.

The industrial importance of steam power was, of course, obvious to the farsighted entrepreneurs who actively developed and invested in it, from Roebuck, Boulton, and the Wilkinsons to the Stephensons. But the larger- and longer-run implications of its use – from 'satanic mills', child labour, and Marxism to acid rain and climate change – were not foreseen at all. Nor did the early steam engineers concern themselves with such questions. They simply sensed an economic opportunity and exploited it. Only in later generations did historians begin to piece together an understanding of the causal connection between the eighteenth century technological changes and the nineteenth and twentieth century economic, social, and environmental consequences.

Whereas the first industrial revolution began in one country (Great Britain) and spread only gradually to France, Germany, the United States, and beyond, today's technology spreads as far and as fast as modern communications networks allow. The scope and speed of the change in physical production technology that is now under way mean that there will be less time for society to react. That means there is less time for decision makers – whether in government or industry – to think about appropriate governmental responses before the imperative to act becomes irresistible.

Moreover, the industrial and political leaders of most countries are more sensitive to public opinion than ever before – another consequence of the new communications technologies. It is not a moment too soon for those in positions of responsibility to anticipate some of the challenges that lie ahead. It is by no means too soon to consider the policy options that are likely to be available – or unavailable – under various plausible future scenarios. (In fact, it may be more important for decision makers, and their advisors, to be aware of the attractive dead ends and seductive potential traps than anything else.)

The purpose of the CIM project [and, indeed, the whole Technology, Economy, and Society (TES) program] at IIASA is to close the widening gap between the pace of technological, economic, and social events, on the one hand, and the progress of understanding those events, on the other.

**The IIASA CIM Study**

The project, which is currently approaching completion, is not the first at IIASA on the application of computers to manufacturing. In fact, one of the first conferences ever held at IIASA (October 1–3, 1973) concerned the automated control of industrial systems. It was co-chaired by Professor A. Cheliustkin of the Institute for Control Problems, Moscow, and Professor I. Lefkowitz of Case Western Reserve University, Cleveland. Topics proposed for discussion at that conference included automation of quality control, economies of scale, increased demand for flexibility, automation of multi-product plants with multipurpose equipment, and adaptation of control systems to changing industrial environments. All of these topics are relevant today, and are discussed at appropriate places in this book.

The IIASA study has attempted, first, to define the existing world situation with regard to the underlying technologies of CIM, and the degree to which technologies such as NC/CNC machine tools, robotics, and CAD/CAM are currently being used in metal products manufacturing. We have concentrated our attention primarily on this subset of the overall manufacturing sector because of two simple facts: *machines* are metal products, and manufacturing (not to mention transportation, mining, agriculture, utilities, and even households) depends on machines. In other words, the metal-products sectors, and particularly the *machine-building* subsectors, have a unique and vital role in the economic system. It is this sector that produces the capital goods on which all production depends. Above all, it is the machine-building sector that reproduces itself.

The methodology adopted in the study is eclectic. It is multi-perspective and multidisciplinary, as well as multinational. It incorporates elements of both 'bottom-up' and 'top-down' approaches. Finally, it incorporates both historical analysis and 'model' forecasts of the future, together with scenario analyses.

The bottom-up part of the study began with a review of the extensive international literature. We have compiled, from (mainly) published sources, a very large worldwide database on so-called flexible manufacturing systems (FMSs). The database contains 880 entries. This database is incomplete in many respects, but it is large enough to carry out a variety of statistical analyses and comparisons, both cross-sectional and longitudinal.

In addition, with help from local collaborating institutions, a number of in-depth case studies at the firm level have been carried out in several countries, with both centrally planned and market economies. The results of this work, when complete, will provide perhaps the first truly international comparison of manufacturing technology and manufacturing management of its kind.

The top-down elements of the study consist of two parts. One is the development of a generalized theory of manufacturing as a process of adding information to materials. In particular, one can regard manufacturing as a process of adding morphological information (i.e., shape and form) to materials. Insights derived from this perspective are utilized throughout the project, although the theory itself is somewhat too technical to be included among the main reports of project results. (It is to be published separately as a scientific monograph entitled *Information, Evolution and Economics*.)

The other part of the work that can be called top-down is analysis using various economic models. One category of economic models that we have used is the static or quasi-static input-output (I-O) type of model to explore international impacts of CIM adoption at the sectoral level. Another type of model used for comparing economic growth scenarios on an international basis is a set of linked macro-models, previously developed by Wilhelm Krelle at the University of Bonn in collaboration with IIASA (the Bonn–IIASA model). Still a third type is the dynamic simulation model with which we are exploring interactions among such variables as R&D expenditure, product improvement, process improvement, capital flexibility, cost reduction, price reduction, demand, capital investment, profitability, and so on.

Outputs of the bottom-up part of the study, in addition to this summary volume (Volume I), will include three major components. The first (Volume II) will be a survey of CIM technologies, present and future. It will include technical aspects of the application of these technologies. The second (Volume III) will be a descriptive and interpretive analysis of the processes of diffusion and adoption of CIM technologies, by technology, sector, and country. Outputs of the top-down part of the work will be presented in a fourth volume (Volume IV), focusing on economic and social impacts of CIM. The first two-thirds of Volume IV will be primarily economic and quantitative in orientation. The final third will be primarily social and qualitative in orientation.

# THE INTERNATIONAL INSTITUTE FOR APPLIED SYSTEMS ANALYSIS

is a nongovernmental research institution, bringing together scientists from around the world to work on problems of common concern. Situated in Laxenburg, Austria, IIASA was founded in October 1972 by the academies of science and equivalent organizations of twelve countries. Its founders gave IIASA a unique position outside national, disciplinary, and institutional boundaries so that it might take the broadest possible view in pursuing its objectives:

*To promote international cooperation* in solving problems arising from social, economic, technological, and environmental change

*To create a network of institutions* in the national member organization countries and elsewhere for joint scientific research

*To develop and formalize systems analysis* and the sciences contributing to it, and promote the use of analytical techniques needed to evaluate and address complex problems

*To inform policy advisors and decision makers* about the potential application of the Institute's work to such problems

The Institute now has national member organizations in the following countries:

**Austria**
The Austrian Academy of Sciences

**Bulgaria**
The National Committee for Applied Systems Analysis and Management

**Canada**
The Canadian Committee for IIASA

**Czech and Slovak Federal Republic**
The Committee for IIASA of the Czech and Slovak Federal Republic

**Finland**
The Finnish Committee for IIASA

**France**
The French Association for the Development of Systems Analysis

**Germany**
Association for the Advancement of IIASA

**Hungary**
The Hungarian Committee for Applied Systems Analysis

**Italy**
The National Research Council (CNR) and the National Commission for Nuclear and Alternative Energy Sources (ENEA)

**Japan**
The Japan Committee for IIASA

**Netherlands**
The Netherlands Organization for Scientific Research (NWO)

**Poland**
The Polish Academy of Sciences

**Sweden**
The Swedish Council for Planning and Coordination of Research (FRN)

**Union of Soviet Socialist Republics**
The Academy of Sciences of the Union of Soviet Socialist Republics

**United States of America**
The American Academy of Arts and Sciences

# Chapter 1

# Overview and Theses

## 1.1 Background: The Barrier and the Breakthrough

The title of this book reflects the subject of the study that it attempts to summarize and synthesize. It makes a nontrivial claim that we are in the early stages of a new industrial revolution, of a magnitude and significance comparable with the "first" industrial revolution that began some two centuries ago. We have identified computer integration (of manufacturing) – i.e., CIM – as the key *breakthrough* in the current industrial revolution. (A more formal definition of this concept is provided in Section 1.4.)

The very notion of a breakthrough presupposes a *barrier* that had to be overcome: a period of progress denied (or slowed to a crawl) owing to some bottleneck or impasse (Ayres, 1988c). It is often the case, where a very big and pervasive change is involved, that the nature of the barrier was not clearly recognized until later. Indeed, it is often the function of historians to perform this recognition and naming function. In the case of the first industrial revolution, historians generally agree that scarcities of charcoal and water power were largely responsible for the intense interest in coal as a source of energy and steam as a source of power in eighteenth century Britain. Is it possible, at this early stage, to identify the corresponding barriers and bottlenecks for the revolution now in progress?

It is already a fairly widely shared assumption – call it "conventional wisdom" – that the computerization of manufacturing results from the convergence of two trends. The most obvious trend is the spectacular technological progress in microelectronics: electronic sensing, electronic data processing, and electronic controls since the end of World War II. This technological factor may be thought of as the "supply side" of the convergence phenomenon.

But *supply* – whether of consumer goods or technology – is, more often than not, called forth by *demand.* In the case of CIM, it is widely assumed, nowadays, that the need (or demand) for computer integration of manufacturing arises from a need for greater flexibility. The inflexibility of mass-production technology, and the so-called hard automation and rigid mode of industrial organization that evolved with it – called "Fordism" or "Taylorism–Fordism" in Europe – is fairly well known. As manufacturing has become more efficient and more capital intensive, over the years, the costs of product change have progressively risen, and the rate of technological innovation in the mass-production sectors, by contrast, has slowed down.

Since computers are by nature programmable, it is generally assumed that the computerization of manufacturing functions is tantamount to introducing greater programmability of production operations. Programmability in this sense is, at least, a prerequisite of flexibility. On deeper analysis, however, the linkages between computerization and flexibility are less solid than they appear to be. On the one hand, the increasing importance of operational (and capital) flexibility cannot be seriously doubted: it is emphasized in every survey of the motives for investment in advanced automation. On the other hand, the proposition that computers increase flexibility *ipso facto* is extremely dubious. (In fact, we have concluded that it is not true.)

Computers are certainly not more flexible than humans. So the flexibility argument – at least in the narrow sense – is only applicable to the substitution of programmable automation for hard automation (e.g., mechanical transfer lines). While there are a few examples of this sort of substitution, the true situation is considerably more complicated. The following abbreviated "mini-thesis" states our view. It is plausible enough to be accepted as a reasonable basis for further discussion and analysis.

## 1.2   The Complexity/Variety Barrier

The imperative demand for ever higher performance has forced products themselves to become increasingly complex and precise. Technological progress in the direction of superior performance has generally been accompanied by increased complexity of both products and processes. It is true that simplifications in products have sometimes followed from improved processes

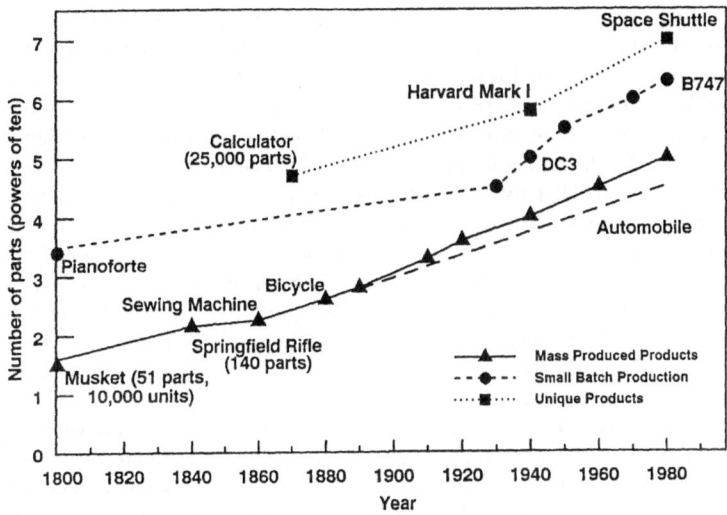

**Figure 1.1.** Complexity trends.

(for instance, the stamped or cast automobile wheel), or in other cases from design simplifications made possible by changes in the philosophy of repair (e.g., the substitution of "snap-ons" for threaded connectors); however, these instances are rather exceptional in the mechanical world. *Figure 1.1* illustrates the historical trend in complexity of products, at least measured in terms of the number of parts.

A Colt revolver or a musket (c. 1850) would have required fewer than 50 parts, all of which could be made in the same armory. An all-metal (brass) Jerome clock of the 1830s would have required fewer than 100 parts, of which about 10 were moderately complex gear wheels and escapements (stamped) and the rest were mostly bolts, nuts, pins, axles, bushings, washers, and flat stamped casing parts. Almost all of these parts were probably made in the same plant. An early sewing machine (c. 1860) would also have required around 100–150 parts, including some stamped parts, several castings, a number of standard items (bolts, nuts, washers, axles, pins, and several gear wheels), and a few complex parts requiring machining or forging (Hounshell, 1984). Some of these parts were probably purchased.

Ball bearings began to replace sleeve bushings in the 1870s and represented a sharp increase in mechanical complexity. They found an important application for the first time in bicycles (c. 1885). This period probably also marks the beginning of the trend toward subcontracting for specialized mechanical components. A bicycle uses 5 or 6 ball bearings, each consisting of 12–20 steel balls rolling between 2 steel races. The bicycle chain alone consists of around 300 individual parts, and the lightweight spoke wheel involves a rather complex hub, an outer rim, and 30–40 spokes with threaded ends plus several fasteners. Altogether, a multi-speed bicycle requires around 800 distinct parts. The typical bicycle manufacturer of today is likely to produce only the welded frame and some key parts like the wheel hubs and derailleurs. Most other parts are purchased from subcontractors, including the ball bearings, nuts and bolts, cables, chains, bushings, gear wheels, tires, and other plastic, glass, or rubber items.

Early automobiles were largely based on bicycle technology, with the addition of a crude internal-combustion engine. A rough estimate for an early motorcar (c. 1900) would be 1,500–2,000 parts, mostly simple adaptations from bicycles or carriages. Later models have become far more complex in almost every way (except for the substitution of stamped metal wheels for bicycle-type spoked wheels). In fact, automobiles at the present time require more than 30,000 distinct parts. Of these, only 10%–15% are produced by the nameplate manufacturer. A modern industrial circuit breaker requires 1,300 parts, while a 1970's IBM Selectric typewriter required 2,700 distinct parts. Roughly speaking, consumer products increased in complexity by a factor of 10–15 from 1830 to 1900 and by another factor of 10–15 from 1900 to 1980.

Increased complexity of products increases the relative importance of assembly in the manufacturing process. It was the difficulty of assembly – then called "fitting" – that first created the need for interchangeability of parts, in the early nineteenth century. This, in turn, resulted in a powerful trend toward standardization of design and manufacturing processes. This trend also resulted in significant economies of scale that enabled manufacturers such as the Singer Sewing Machine Co. (and, most spectacularly, Ford Motor Co.) to cut their costs very sharply toward the end of the nineteenth century and in the first decades of this century. (Examples of the magnitudes of manufacturing productivity increases from the 1840s to 1897 are shown in *Table 1.1*.)

The problems of large-scale production of standardized products led to an organizational structure and philosophy of extreme specialization and

**Table 1.1.** Productivity increases: 1836 to 1897.

| Item | Period | Increased output per man-hour (multiplier) |
|---|---|---|
| *Metal Products* | | |
| Pitchforks (steel) | 1836–1896 | 15.60 |
| Plows, iron, and wood | 1836–1896 | 3.15 |
| Rakes, steel | 1858–1896 | 5.96 |
| Axle nuts (2″) | 1850–1895 | 148.00 |
| Carriage axles | 1856–1896 | 6.23 |
| Carriage axles (4″ steel) | 1862–1896 | 6.23 |
| Tire bolts (1 3/4″ x 3/16″) | 1856–1896 | 46.90 |
| Carriage wheels (3′6″) | 1860–1895 | 8.41 |
| Clocks, 8-day brass | 1850–1896 | 8.30 |
| Watch movements, brass | 1850–1896 | 35.50 |
| Shears, 8″ | 1854–1895 | 5.51 |
| Saw files, 4″ tapered | 1872–1895 | 5.51 |
| Rifle barrels, 34 1/2″ | 1856–1896 | 26.20 |
| Welded iron pipe, 4″ | 1835–1895 | 17.60 |
| Nails, horseshoe, no. 7 | 1864–1896 | 23.80 |
| Sewing machine needles | 1844–1895 | 6.70 |
| *Other Products* | | |
| Bookbinding, cloth (320 pp) | 1862–1895 | 3.80 |
| Mens shoes, cheap | 1859–1895 | 932.00 |
| Womans shoes, cheap | 1858–1895 | 12.80 |
| Hat boxes, paperboard | 1860–1896 | 3.22 |
| Wood boxes (18″ x 16″ x 9″) | 1860–1896 | 9.73 |
| Paving bricks | 1830–1896 | 3.89 |
| Buttons, bone | 1842–1895 | 4.04 |
| Carpet, Brussels | 1850–1895 | 7.95 |
| Overalls, men's | 1870–1895 | 10.10 |
| Rope, hemp | 1870–1895 | 9.74 |
| Sheet, cotton | 1860–1896 | 106.00 |
| Electrotype plates | 1865–1895 | 2.91 |
| Chairs, maple | 1845–1897 | 6.43 |

Source: Ayres (1984), data from US Department of Labor.

division of labor that was relatively unsuited to small-scale production of nonstandardized or customized products. It was also inflexible in the sense of being very capital intensive and thus poorly adapted to respond to unexpected changes in market conditions or technology. Thus, the biggest mass producers (mainly the US auto manufacturers) were relatively comfortable as oligopolists in an environment where they had a substantial degree of control over demand (via advertising). They were correspondingly less adaptable to external events such as the "oil shock" of 1973–1974.

As personal disposable incomes have increased with general economic growth and markets have globalized, demands for product variety and "customization" have increased much faster than demands for standardized products. In effect, customers are prepared to pay a significant premium price for greater choice. Manufacturers in most countries, facing much stronger foreign competition in the 1980s than they did in the early 1970s, have no real choice but to satisfy that demand by offering greater choice. However, variety does not come at zero cost, since variety can only be achieved by sacrificing economies of scale, at least to some extent. The "volume-variety trade-off" is illustrated in *Figure 1.2*.

When the large number of different models of complex modern products is considered, the problem of organizing production (and subsequent service) becomes truly staggering. A major manufacturer of electrical connectors (AMP) produces 80,000 different types. IBM's Selectric typewriter was made in 55,000 different models. Westinghouse Electric Co. (c. 1983) manufactured over 50,000 different turbine wheel shapes for its steam turbines. Caterpillar Tractor Co. (c. 1985) had over 25,000 different subcontractors making various component parts of its machinery products. There are more than 1,900 automobile models made in the world today.

The so-called major manufacturers have to a large extent become "systems integrators," providing only some of the more specialized parts and final assembly of subsystems from a network of suppliers. Their major economic role is design, marketing, and service, not production *per se*. For such firms, direct manufacturing labor constitutes a minor proportion of all costs, ranging from 15% to 25% in some of the mature manufacturing industries to 5% or even less in the case of some high-tech firms such as IBM. A pattern that holds for the most successful firms today is likely to be true of the majority of firms in the future.

It is sometimes argued that the product *life cycle* is getting shorter. The main evidence for this seems to be recent experience in the semiconductor industry (discussed in more detail in Chapters 4 and 5), where major new

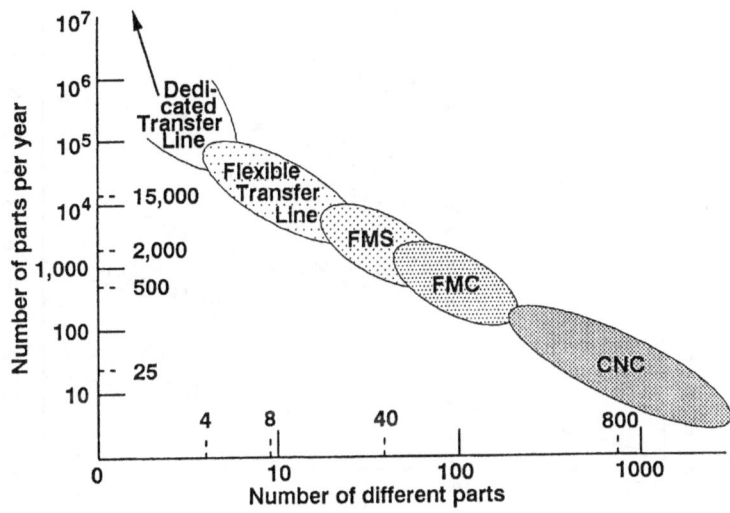

**Figure 1.2.** Volume-variety trade-off.

products are now introduced roughly every 3 years, as compared with model cycles of 5 to 7 years or more in past decades.[1] If this acceleration of product innovation applies across the whole range of manufacturing industries, it would partially account for the increased concern with organizational and capital flexibility in production.

The combined impact of increasing product *complexity*, together with increased *variety*, on manufacturers has been to create a massive problem of information management and coordination. The response of most firms has been to computerize many individual functions *ad hoc* without serious planning for their interaction with each other. Most large organizations now find themselves with hundreds, if not thousands, of incompatible programs and computers that cannot communicate directly with each other. But they all generate reams of printout and generate enormous data requirements that impose unexpected loads and costs on organizations.

As a direct consequence of this trend toward increasing complexity-plus-variety, information-based activities now constitute a growing fraction of all activities, both within individual manufacturing firms and in the economy

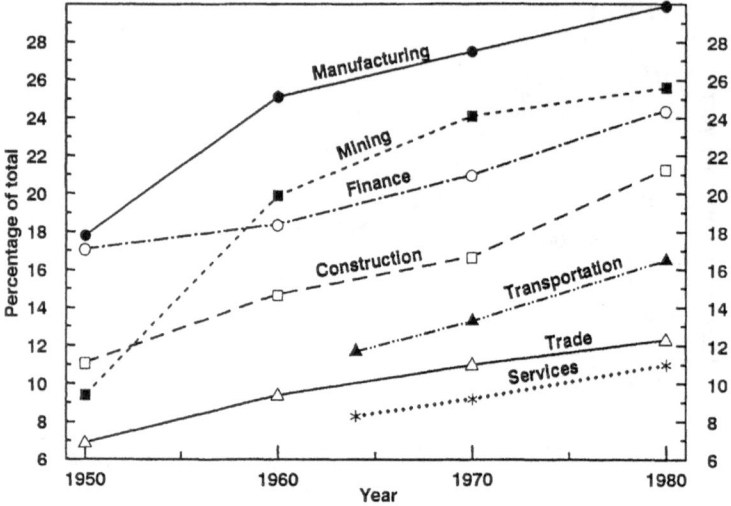

**Figure 1.3.** Nonproduction workers as percentage of total employment by major sector: USA. Source: *Employment & Training Report of the President* 1981, p. 213.

as a whole. More and more people are employed in *overhead* activities, as shown in *Figure 1.3*. A substantial, and growing, fraction of manufacturing jobs are information-related.

As a matter of interest (which is considered in more detail in Chapter 7), the postwar rate of productivity growth for the OECD economies slowed down sharply after 1975 (*Table 1.2*). It is possible that at least part of this slowdown may be attributable to a buildup of problems of *information overload* due to haphazard and *ad hoc* computerization of many specific tasks without adequate provision for upstream and downstream consequences, including interference phenomena.

The interference problem is particularly acute when rapid computerization at the individual task or functional level occurs within the rigid Taylorist–Fordist organizational structure, leaving most of the *information transducer* (input–output) functions for human workers. The problem seems to arise from the fact that the information output of the computerized

Robert U. Ayres    9

**Table 1.2.** The productivity slowdown (1973–1985): Average rate of growth in total factor productivity (TFP) in period.

| Country | Pre-1973 | Post-1973 | Difference |
|---|---|---|---|
| Japan[a] | 5.97 | 1.19 | −4.78 |
| FRG[a] | 2.78 | 0.83 | −1.95 |
| France[a] | 2.27 | 0.32 | −1.95 |
| Italy[a] | 2.67 | 0.27 | −2.40 |
| Netherlands[a] | 1.77 | −0.46 | −2.23 |
| Austria[a] | 2.62 | 1.50 | −1.12 |
| USA[b] | 2.09 | 0.55 | −1.54 |
| USA[c] | 2.48 | 0.36 | −2.12 |
| USA[d] | 2.30 | 0.50 | −1.80 |
| USA[e] | 1.79 | 0.71 | −1.08 |

[a]OECD 1950–1973, 1973–1985.
[b]Dennison (A) 1948–1973, 1973–1979.
[c]Dennison (B) 1948–1973, 1973–1987.
[d]Kendrick 1948–1973, 1973–1987.
[e]Jorgenson 1948–1973, 1973–1979.
All data and complete sources taken from Abramowitz, 1989.

functions can easily overload the human transducers, resulting in an increasing amount of "garbage information" (errors) or simply *noise* in the system, which is costly to detect and correct.

Another impact of increasing complexity and variety on manufacturers, whose extent has not been widely recognized yet, has been to create an acute problem of quality control. Inspection, rework, and after-sales service account for an increasing fraction of total costs – sometimes exceeding 60%. These costs could be eliminated, in principle, if there were no errors or defects in the manufacturing process. Yet it is well known to industrial engineers that most product defects result from human errors, either in design and engineering or in fabrication. (Errors resulting from mechanical failures are comparatively infrequent, and comparatively easy to identify.)

The human propensity to make errors is inherent. In fact, it is plausible to suggest that the human propensity for making errors, resulting in defective parts and eventually defective products, is the *barrier* noted in the second paragraph of Section 1.1 (Ayres, 1988b). This propensity can surely be reduced by sophisticated ergonomics: training, motivation, and sophisticated work place design. But "to err is human." The probability of error cannot be eliminated, or even reduced below some physiological-psychological limit. To be sure, human design/engineering errors can, in principle, be very sharply reduced by the use of computers for assistance in those processes (CAD/CAE). Errors in machine operations and assembly can also by reduced sharply by using computers for controlling the manufacturing process (CAM), and ultimately can be cut virtually to zero as human workers are phased out of on-line functions. ˒

An implication of the above is that the real motivation for introducing CIM is to simultaneously achieve both flexibility (to respond quickly to market signals) and improve quality control. Consumers in the wealthier countries are increasingly insisting on combinations of variety and quality that are more and more out of reach to traditional (i.e., manual or mechanized, but not computerized) manufacturing technology. This trend means that it is becoming more and more difficult for a manufacturer based in a Third World country to exploit low labor costs as a basis of competitive advantage in OECD markets, at least for complex products such as automobiles, aircraft, and capital goods. Manufacturers without CIM and all it implies will not be able to achieve the levels of product quality that will be demanded. (The sad fate of the Yugo automobile in the US market illustrates the problem.)

A further implication of this reasoning is that the export-based economic development strategy adopted by Japan, and successfully imitated (to a point) by the "Four Tigers" of East Asia, is not likely to be usable in the future by countries like Brazil, Mexico, India, and China. Even for Taiwan, the ability to manufacture bicycles and personal computers using traditional labor-intensive means may not translate into an ability to make automobiles, aircraft, and sophisticated computer or telecommunications systems. Rather, small- and medium-sized firms in such countries may be effectively limited for a long time to producing the less complex consumer products or subassemblies (such as printers or keyboards) or both as subcontractors to big firms based in Japan, Europe, or North America. To be fully competitive across the whole range of sophisticated manufactured products will necessitate a commitment to capital-intensive, computer-controlled, virtually unmanned production facilities that will be very difficult to justify politically in countries with large pools of unskilled labor.

## 1.3   Globalization and Restructuring of Industry

The trends toward increasing product complexity and variety have been in progress for two centuries. Why have these difficulties, which are not new,

**Table 1.3.** Tariffs on automobiles as percentage of customs value.

| Year | USA | Japan | France | Germany (FRG as of 1950) | Italy | UK |
|------|-----|-------|--------|--------------------------|-------|-----|
| 1913 | 45 | n.a. | 9–14 | 3 | 4–6 | 0 |
| 1924 | 25–50 | n.a. | 45–180 | 13 | 6–11 | 33.3 |
| 1929 | 10(1930) | 50 | 45 | 20 | 6–11 | 33.3 |
| 1932 | 10 | n.a. | 45–70 | 25 | 18–123 | 33.3 |
| 1937 | 10 | 70(1940) | 47–74 | 40 | 101–111 | 33.3 |
| 1950 | 10 | 40 | 35 | 35 | 35 | 33.3 |
| 1960 | 8.5 | 35–40 | 30 | 13–16 | 31.5–40.5 | 30.0 |
| 1968 | 5.5 | 30 | 0–17.6 | 0–17.6 | 0–17.6 | 17.6 |
| 1973 | 3.0 | 6.4 | 0–10.9 | 0–10.9 | 0–10.9 | 10.9 |
| 1978 | 3.0 | 0 | 0–10.9 | 0–10.9 | 0–10.9 | 0–10.9 |
| 1983 | 2.8 | 0 | 0–10.5 | 0–10.5 | 0–10.5 | 0–10.5 |

n.a. = not applicable.
Source: Altschuler *et al.*, 1984 (Table 2.2, p.17).

finally emerged as a major barrier to progress? Some other factor or factors must be involved.

Changing industrial structure and globalization of markets are partially responsible for the new international situation. For many reasons that cannot be fully detailed here, the world economy has become more and more open and competitive in the last two decades. As a result of successive international agreements (known as the General Agreement on Tariffs and Trade, or GATT), tariffs on most goods have decreased sharply since the 1950s, and non-tariff barriers (NTBs) have been reduced considerably in many countries (see *Table 1.3*).

Globalization of markets has had another consequence that may not have been fully appreciated as yet. As recently as the 1960s the United States (for one) was nearly self-sufficient with regard to manufactured goods, and many less wealthy countries such as India and China could reasonably aspire to self-sufficiency, albeit on a less affluent scale. Over the past two decades, increasing competition on a global scale has made self-sufficiency an idea whose time is past. Even the (still) vast US manufacturing establishment is now incapable of making all the manufactured products it needs.

For example, more than 90% of the commercially marketed dynamic random-access memory (DRAM) chips needed for computers and other products are made in Japan. Japan also dominates the world production of cameras, watches, radios, TVs, and VCRs. Japan, on the other hand, must import most of its microprocessors, CAD systems, commercial airliners, and large jet engines. (In fact, there are only three manufacturers of such engines in the Western world: GE and Pratt & Whitney in the USA and Rolls-Royce in the UK.) Other examples of products made by only a few specialized firms include large telephone-switching systems, supercomputers, large civil aircraft, helicopters, large steam turbines, large diesel engines, large earth-moving machines, heavy mining equipment, rolling mills, large ships, and oil drilling equipment.

Except for Japan, West Germany, and the USA, most countries in the Western world are totally dependent on external sources of supply for some, if not most, of these – and many other – goods. For smaller countries, of course, self-sufficiency in manufacturing was never a realistic possibility. Switzerland makes no computers, cars, or aircraft, but it sells large steam turbines, generators, diesel engines, pharmaceuticals, and watches. Sweden is strong in the metal-working sectors, but it has virtually no chemical industry and limited electronics, and so on. International trade is no longer a luxury; it is a necessity for survival.

In Eastern Europe the regional specialization (in response to increasing product complexity) became even more extreme. This is because in Eastern Europe central planning authorities concentrated on maximizing the benefits of economies of scale. As a result, there are essentially no small firms. Markets have been allocated for many years by a central planning organization (Council for Mutual Economic Assistance, CMEA) on the general principle that each country specializes in certain groups of products. The USSR sells (or barters) petroleum and natural gas to its trading partners in exchange for machinery and manufactured goods. Only the USSR (by far the largest) is even remotely self-sufficient. Within the USSR, regional specialization is also the pattern. For example, it was recently revealed that the troubled republic of Azerbaijan now produces only 3% of the petroleum output of the USSR, but it accounts for two-thirds of the oil-service industry in that country.

As protectionist barriers have fallen, markets have become bigger and more competitive. There is no longer a French domestic market, a German domestic market, or even a US domestic market. Multinational firms have branches or partners everywhere. Export-led growth on the Japanese model has become a feasible strategy for economic development in the Pacific Rim countries (South Korea, Taiwan, Hong Kong, Singapore, Malaysia, and Thailand). The export-oriented manufacturing firms located in countries such as Switzerland and Sweden and the Benelux countries, which were never self-sufficient, have been better able and quicker to adapt to this change than the domestic-oriented firms in the USA. This is clearly reflected in the declining share of US-based firms in the US market, from steel and automobiles to electrical equipment and "high technology."

Whether as a consequence of globalization of markets or other factors, firms faced with declining domestic profit opportunities have typically reacted either by investing more abroad in faster-growing economies or by turning to the stock market and trying to grow faster by acquisition of smaller firms. The accelerated foreign investment by US-based firms – mainly in the late 1960s and 1970s – undoubtedly helped Western Europe narrow its productivity gap with the USA to the point where there is no longer much difference.

To the extent that firms began to invest in the shares of other firms rather than in new plants and equipment, a new sort of industrial organization was created – the diversified conglomerate. There was a major wave of conglomerate mergers in the USA in the late 1960s and early 1970s. More recently, many US conglomerates created 20 years ago have become targets of restructuring by another wave of financial manipulators, who see non-performing assets that can be spun off to management groups or to other firms. At the same time, however, the number of mergers and acquisitions has reached an all-time peak, and new conglomerates are being created even as older ones are being dissolved.

In Europe, on the other hand, the gradual solidification of the EEC as an effective entity has created new and enticing prospects for mergers and cooperation across national boundaries. Firms are growing in size (by merger) to exploit larger markets, though usually they stay in the same line of business. Deregulation and privatization of some traditional state monopolies have also created new and exciting opportunities and attracted investment. The new access to the economies of Eastern Europe and China seems to promise both opportunities and dangers for the future. The present is a time of great potential for change, and of great uncertainty.

## 1.4 CIM and the Systems View of Manufacturing

As stated at the outset, this book is dedicated to the proposition that CIM is revolutionary. In the words of Merchant (1989a) one of the pioneers of the CIM concept, "it is a *wholly new* approach to the operation of manufacturing in its entirety...that...offers *enormous* potential to improve manufacturing capability and cost-effectiveness."

Having briefly reviewed some of the background, it is perhaps appropriate to insert a more formal definition of CIM. This widely used acronym stands for Computer Integrated Manufacturing. These words mean different things to different people. Many explicit definitions have been offered – too many to summarize. Here is one, from Digital Equipment Corporation (DEC):

> CIM is the application of computer science technology to the enterprise of manufacturing in order to provide the *right* information to the *right* place at the *right* time, which enables the achievement of its product, process and business goals.

What is important for purposes of this book and its sequelae is that the term CIM is used here in the broadest possible sense. While much of the discussion hereafter deals with activities and technologies found on the factory floor, the key to CIM is *integration* of manufacturing and office functions, both off-line and on-line.

This concept dates back to the early 1960s (Merchant, 1962). It was a natural consequence of the emerging realization (attributable, in part, to the new modes of thinking associated with the digital computer) that manufacturing is, indeed, a *system*, not merely a collection of unrelated activities linked by a common end product or a common corporate owner. What differentiates a "system" (in the technical sense of the word) from such a collection is the *interdependence* of the components. *Figure 1.4* displays the systems view of CIM in a diagrammatic form (Merchant, 1989a). The importance of interdependence is clear from the diagram, although the relationships are oversimplified.

It has been said (of the natural environment) that "everything depends on everything else." This may not be literally true of a manufacturing system, but it is far more apposite than the converse. Yet it is the converse proposition – that every activity in manufacturing is independent of every other activity – that underlies the guiding philosophy of most industrial management during most of this century, viz., "Taylorism." Thus, in a very fundamental sense, CIM is antithetical to, and incompatible with, Taylorism, specialization, and the division of labor (see Chapter 7).

The systems view of manufacturing has been elaborated, in particular, by Joseph Harrington (1984) in his influential book *Understanding the Manufacturing Process*. The real complexity of the interdependences between the various functions is illustrated in *Figures 1.5* and *1.6*. A clear and provable

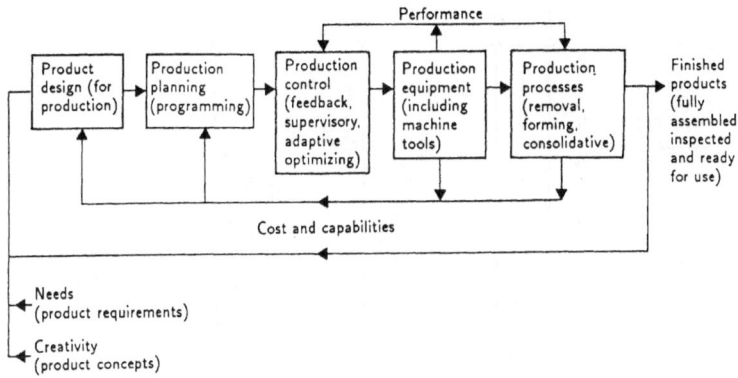

**Figure 1.4.** Concept of the computer integrated manufacturing system: 1969. Source: Merchant, 1989a.

implication of this interdependence is that optimum performance of the system as a whole is not achievable, in principle, by maximization of output at the component level.

Another important consequence of the systems (or holistic) view is that extreme specialization of function (and of labor) is counterproductive. A CIM system cannot be implemented with "old style" division of labor or "old style" hierarchical management. These points have been repeatedly cited as conclusions drawn from experience. They are so fundamental that they can virtually be regarded as axiomatic.

CIM also has a critical technological component: namely, the increasingly explicit (and increasingly digitized and computerized) knowledge base underlying many of the processes and the substitution of explicit decision algorithms, implemented on electronic computers, for many judgmental (and error-prone) "microscale" decision processes formerly made exclusively by humans. In brief, the implementation of CIM involves substituting computers for humans in the vast majority of repetitive on-line activities, while retaining humans in those non-repetitive functions (such as conceptual design, planning, problem diagnosis, and "trouble shooting") where human capabilities are still irreplaceable and probably will be for the foreseeable future.

To summarize, CIM is a philosophy of management, predicated on the systems view of manufacturing, which emphasizes integration of functions,

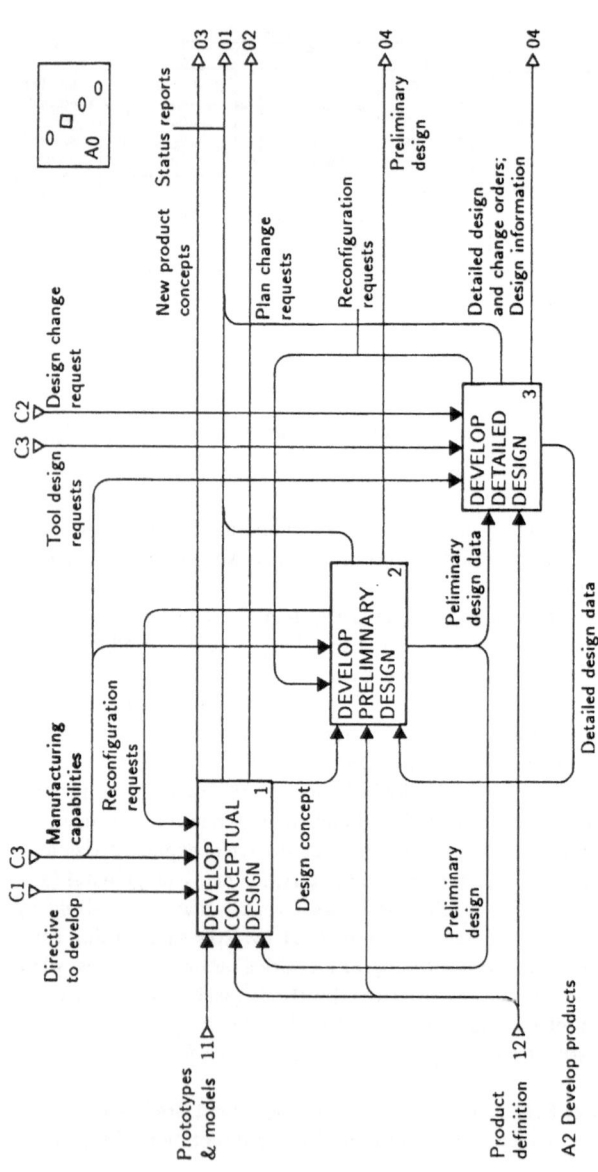

**Figure 1.5.** Product development.

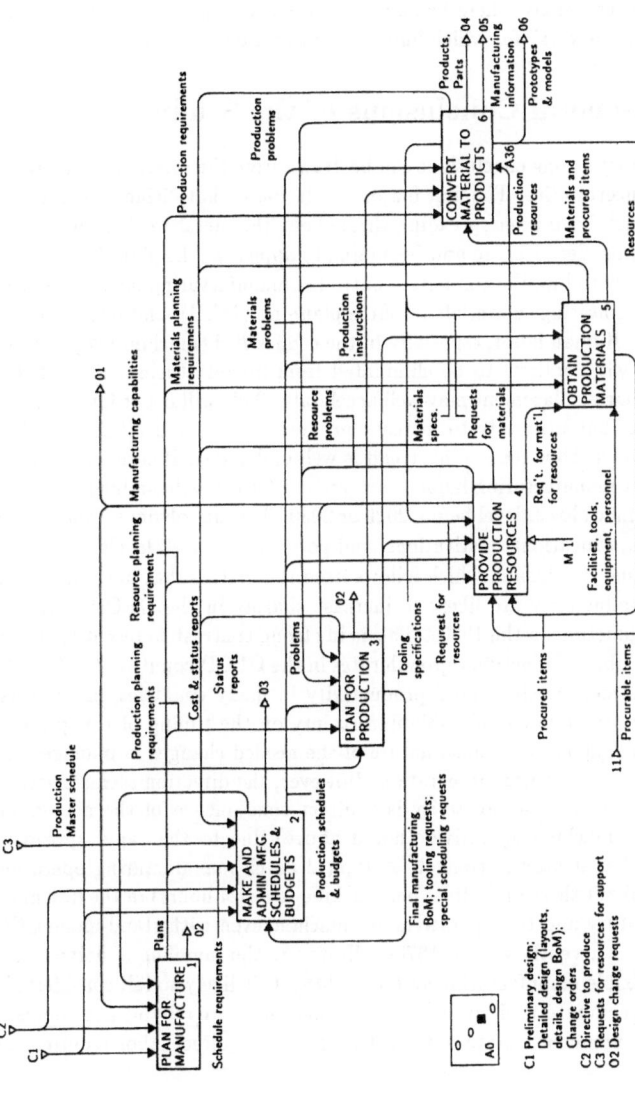

**Figure 1.6.** Product manufacturing. The six-component functions produce a crowded diagram, reflecting the many and complex interactions of the six functions.

flexible multi-skilled labor, and maximum feasible use of computers and formal decision algorithms to replace error-prone humans in repetitive on-line tasks (especially where humans have been used as "information input-output transducers" for which human abilities are inherently ill-suited).

## 1.5    General Conclusions of the Study

Several fairly strong conclusions can be drawn from the study, notwithstanding the uncertainties. The first major conclusion is that "islands of automation," which have been spreading throughout the metal-working industries for over three decades, are now beginning to appear at the plant level, where it is possible to link the computer-controlled manufacturing operations with design, engineering, materials-resource planning (MRP), and other "office" activities. Human labor, especially in the semiskilled (machine) "operative" category, will continue to be eliminated from manufacturing. In fact, the rate of labor displacement may well accelerate. This will occur for a number of reasons, but primarily to increase product quality and reliability while cutting direct labor costs. This trend is well under way. It seems quite clear that direct manufacturing labor – now around 20% of value-added – will decline to a much lower level before 2020 or 2030. This has obvious implications for unions, educational institutions, and government at all levels.

A second conclusion, which follows from the first, is that both labor and capital productivity are likely to increase sharply in the OECD countries starting sometime in the 1990s. This is in strong contrast to recent trends of declining labor productivity growth rates in the OECD countries (*Table 1.2*) and actual declines in capital productivity in many countries and sectors. To be sure there is considerable uncertainty on the timing of this productivity surge, given the radical nature of the needed changes in management philosophy and software integration. However, the direction seems clear.

Increased labor productivity is a direct consequence of the decreasing number of machine operatives (noted above) due to the rapid spread of computer-based process control. A typical batch manufacturing operation required about three operatives per machine (over 24 hours) in the late nineteenth century and two operatives per machine even with stand-alone CNC machines as recently as the 1970s. However, this number is halved in a typical flexible manufacturing system (FMS). It is likely to fall considerably further as computerized machine controls become more reliable and sophisticated. In summary, a state-of-the-art FMS can cut direct labor requirement

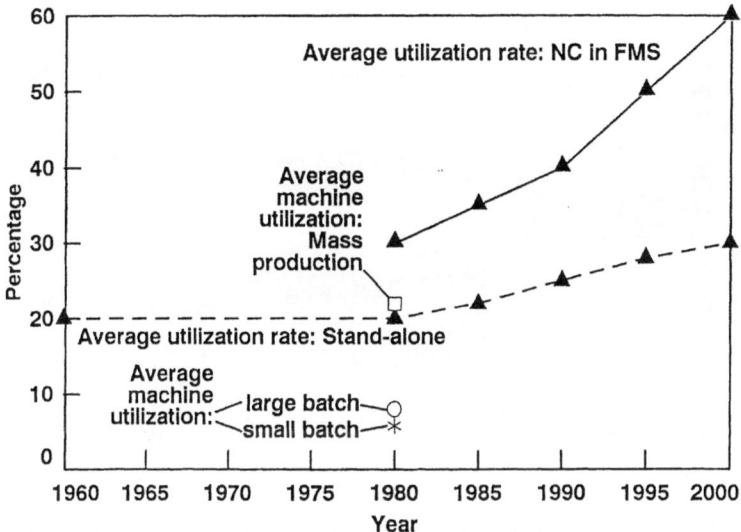

**Figure 1.7.** The impact of computer control on machine utilization.

by 80%–90% as compared with a conventional machine shop, and by 50% as compared with a group of stand-alone NC machines. As CIM technology takes over at the plant level, direct labor requirements will fall in roughly the same proportions for other activities, including assembly.

Increased capital productivity will be a consequence of higher machine utilization, sharply reduced inventories (90% reduction), and smaller factory buildings. *Figure 1.7* shows that even stand-alone NC and CNC machines are utilized at a far higher rate than machines controlled manually. In a typical FMS (c. 1980) average machine utilization (30%, based on a 24-hour day, seven-day week) was half again higher than for similar stand-alone NC machines. This figure should double (to 60%) for state-of-the-art systems by the year 2000. The average for all manufacturing industry lags the leading-edge systems by a decade or so, but even the lag-time seems likely to decrease.

A closely related trend is the increase in quality of production and the decrease in *quality spread*, both resulting from the introduction of CIM. This trend is shown qualitatively in *Figure 1.8*. Improved product quality

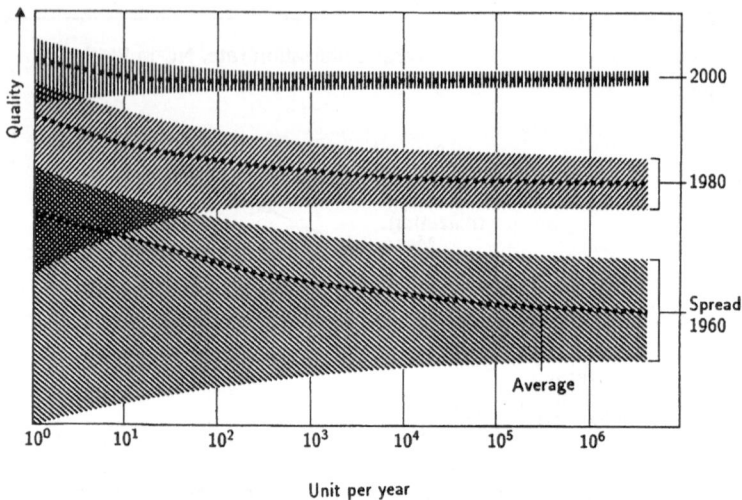

**Figure 1.8.** Increase in quality and reduction of quality spread made possible by flexible manufacturing. Source: Wyss, 1985.

at the level of the firm increases competitiveness, and is consequently a major driving force behind the trend toward CIM. It also has significant downstream benefits to consumers, in terms of extending product life and reducing maintenance costs.[2]

The third major conclusion, which follows from the above, is that productivity gains in manufacturing, especially in the capital goods sectors, will have "ripple" effects throughout the economy. This ripple effect will be felt by other productive sectors and final consumers in terms of lower prices for capital goods and – ultimately – all other goods and services. These indirect effects can be traced by means of input-output models (see Chapter 7). Lower prices, in turn, stimulate consumer demand and thus contribute to further economic (GNP) growth.

The fourth major conclusion of the study follows from the fact that capital goods "hardware" (i.e., machine tools) are becoming more and more standardized. As a result, competitiveness in manufacturing industry will increasingly depend on the quality of a firm's organizational structure and "know-how," as embodied in software. Software engineering (and software security) will become core functions for a world-class manufacturing firm.

Because competitive advantage resides in software, it is far too important to rely exclusively on outside consulting firms or software developers. The critical elements of software, including the basic system architecture, must be developed "in-house." Security will also become a far more complex problem in view of the ease of transferability of software, not to mention the problems of software contamination by computer "viruses."

The importance of "do-it-yourself" has apparently been understood much better by Japanese and West German firms than by US-based firms. Whereas only about 10% of Japanese FMS systems were bought from outside vendors, in the USA the corresponding figure was 70%. This seems to reflect a much stronger commitment to manufacturing excellence on the part of the Japanese. It helps to explain their competitive success in world markets.

A fifth major conclusion of the study, related to the third, is that it is impossible to convert a traditional Taylorist–Fordist manufacturing plant to CIM by simply installing sophisticated new equipment and "patching" incompatible software subsystems together *ad hoc*. There are two fundamental problems. First, the software basis for computer integration cannot be created by incremental additions to what exists now in most firms. It requires a comprehensive "top-down" approach to the problem of communication between different machines.[3] A completely new *meta-system* is needed, with the capability of accommodating continuous change, and eventually learning and evolving – i.e., developing its own "artificial intelligence." This is a tremendous challenge to management. On the one hand, the longer a firm waits to confront this major internal effort, the more difficult it is likely to be, because more existing software will have to be trashed. This fact has received almost no attention in the management literature. On the other hand, there are major risks in trying to computerize a manufacturing system that has not been sufficiently well-understood and rationalized. It is vitally important to simplify all operations as much as possible – starting with design – before investing heavily in software. It is also important to design the integrated system to take full advantage of the capabilities of computers where computers have major advantages (as in computational speed and reliability) while keeping humans involved just in those areas where human capabilities are truly superior. CIM software must, in a sense, take over many functions of the central nervous system of the organism (the firm), leaving only the external intelligence, maintenance, repair, design, R&D, and strategic decision-making functions to the human employees. The division of labor between humans and computer is so important that a successful CIM system can only be implemented with active and intensive involvement by the highest level of management.

The second problem for CIM systems programming and software design is that the off-line supervisory functions that humans must still perform in a CIM plant become much more complex than was the case with earlier generations of stand-alone or mechanically linked machines. This is because the range of normal behaviors of the more complex systems is much larger, and the complex electronic linkages also introduce nonlinearities and associated complex pathologies. In effect, the operators must not only have higher levels of skill than before, but must also be more flexible and capable of doing more jobs. This, in turn, requires a new and more flexible type of organization. It has even been suggested (see Chapter 6) that the major gains of CIM up to today (where it has been effective at all) arise more from the associated organizational reforms than from the new capabilities of the equipment, *per se.*

In this connection it has also been suggested (Chapter 8) that a new type of non-hierarchical *network organization* with *distributed intelligence* and decentralized decision-making powers may be evolving. It can be argued that advanced telecommunications and computer technologies make it possible for functions in different organizations to interact with each other horizontally without necessarily going through all the vertical hierarchical levels. At any rate, it does seem to be true that too many hierarchical levels inhibit vertical information feedback, and thus make it difficult for the higher levels of management to know all they should know before making decisions. It is difficult to anticipate the exact form a future network organization might take, but it is easy to see that a change along these lines would have an enormous impact on the lives of most workers. A possible "scenario" that we did not attempt to evaluate explicitly, but which is consistent with all of the factors driving CIM, is that computerized telecommunication networks linking small firms with specialized marketers can effectively eliminate the economies of scale that large firms currently enjoy in this area. The Japanese trading companies and the Italian textile industry, among others, seem to exemplify this possibility. It would seem to follow that large manufacturing firms may not have much, if any, competitive advantage over small firms in the coming decades, except perhaps in terms of access to specialized engineering expertise.

A sixth conclusion, which follows from the previous one, is that the *software* component of capital will continue to grow in importance *vis-à-vis* the *hardware* component (*Figure 1.9*). The electronic hardware component of

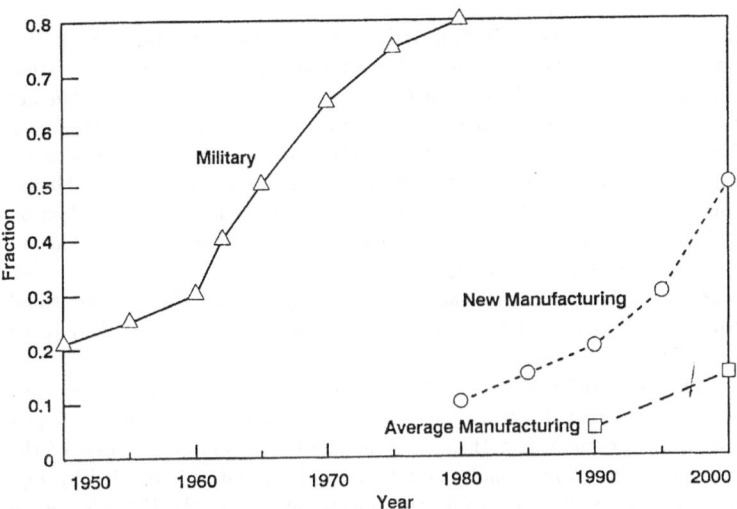

**Figure 1.9.** Software as fraction of total capital investment.

capital stock (computers and electronic controls), which grew rapidly in the 1960s and 1970s, will not continue to grow so fast, because of more sharply declining prices. In fact, by the year 2000 software is likely to be so important that it will have to be explicitly measured. While no such measures presently exist in the national accounting system or the SIC, some indicators are available. It is now a widely accepted rule of thumb that the ratio of software to hardware costs averages around 3:1 for any newly computerized system. This is roughly the reverse of the rule of thumb in the early 1960s. Issues of software inflexibility, software incompatibility, and software productivity are now becoming dominant considerations in designing major systems. An increasingly important objective of computer science research will be the development of *intelligent* (i.e., *adaptive*) programs and software to generate CIM software.

More speculative is the seventh conclusion, which concerns the north–south economic competition. Recent trends indicate a fairly rapid movement of manufacturing away from the high-wage industrialized countries, especially to the perimeter of Asia. This has been particularly noteworthy in the area of electronics assembly and garment manufacturing. It would seem, however, that as the direct manufacturing component of total cost declines, large firms will be increasingly disinclined to fragment their operations in this way, with the accompanying penalties in terms of more complicated logistics, inventory controls, and so on. It is also argued that *economies of scale* are being displaced by *economies of scope*, arising from flexible automation (see Chapters 3 and 7). The logic of the situation would seem to indicate a trend back toward the collocation of production with major markets. Flexible automation seems to reduce the benefits of extremely large-scale production facilities (dictated, in the past, by the costs of "hard" automation). This, in turn, suggests a more dispersed, decentralized production system with many more small plants, located near markets. In other words, the apparent trend toward offshore assembly, especially for US-based firms, seems likely to be reversed.

The competitive advantage of low-wage countries may also be diminished to the extent that, by depending more on human labor than the developed countries, they may find themselves unable to mass-produce goods of the requisite international quality standards. (This problem is implied by *Figure 1.8*.) Thus, it seems likely that increasingly after the 1990s low-wage countries will have only limited access to the markets for manufactured goods in the wealthier countries, primarily at the low end of the quality spectrum. On the other hand, demand for high-quality hand-made goods will surely increase in the industrialized world. The prices of such goods as hand-made carpets and furnishings will continue to rise, and this segment of the manufacturing sector may grow in importance if the necessary skills are not lost.

Beyond these major propositions, a number of possible, but still problematic, corollary propositions and implications lurk. The following is a partial list of additional research hypotheses that can be found in the literature (sometimes stated as facts) or derived from our own work. They cannot be fully resolved at this time; they are simply listed here without extended comment.

It is argued that the *logistics* function (as contrasted with the *production* function) is becoming more important, possibly due to the globalization of technology and markets, combined with increasing specialization of functions. If so, the opportunities for using advanced information technologies to facilitate logistic functions will increase sharply.

On the other hand, the declining importance of economies of scale, noted above, suggests less, rather than more, centralization of production. This would seem to contradict the thesis that logistics are becoming more important because of increasing specialization. Time will tell.

It is argued by some economists that a new "economic paradigm" is evolving. However, economies of scope are not necessarily as efficient as economies of scale, as a mechanism for driving economic growth (Ayres and Zuscovitch, 1990). If there is no mechanism in the "new economic paradigm" that brings down production costs consistently as a function of experience or "learning by doing," there will be no mechanism to increase the size of the potential market (see Chapter 7). In this case, the historical function of scale economies as an engine of aggregate economic growth may result in a long-term slowdown in productivity growth. It is even conceivable that this process has started already and could be a partial explanation of the slowdown in productivity growth since 1973. It could also provide part of the technical explanation for the current economic crisis in the centrally planned countries of Eastern Europe.

Several trends noted above seem to imply a period of great opportunity, accompanied by great risks, for small- and medium-sized manufacturing firms (see Chapter 8). The creation of semipermanent networks among the first, second, and third "tier" of firms (systems integrators being at the top level), will provide greater security for those smaller firms included in the network, and much greater insecurity for those left out. Among the latter, indeed, a wave of failures in coming years can be anticipated. Unfortunately, the first to adopt programmable technologies may not be guaranteed to survive, although the late adopters are almost sure to fail. Timing is extraordinarily critical in making this transition.

Furthermore, if the trend toward unmanned manufacturing is indeed real, there will not be many people working in the manufacturing sector (at least in the advanced countries) 50 years from now (Chapter 8).[4] The analogy with agriculture is compelling. Nobody argues that food is unimportant, but it is a fact that few people are needed to produce enough of it to feed the rest. Agricultural employment has been declining for more than a century (*Figure 1.10*). Similarly, manufactured goods are surely important, and capital goods likewise, but the proportion of the work force directly involved in industrial production in the USA has been declining (in percentage terms) since 1950, and the trend continues. By 2020 the fraction may be comparatively small. There will be a noticeable impact on the economic power of labor unions (and hence on the advantages of union membership). "Work place" issues will increasingly be left to the political domain.

A final question then arises: Does the advent of CIM imply massive unemployment? If not, what will the former manufacturing work force be doing? Will the middle class decline? A detailed answer must be based on

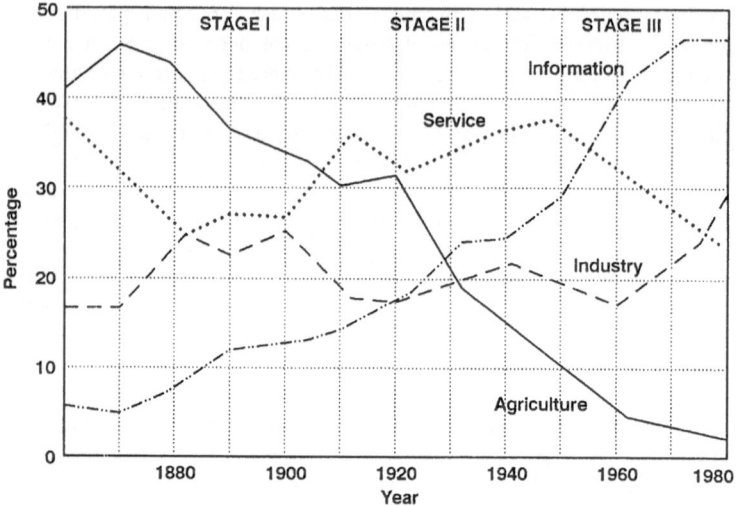

**Figure 1.10.** Trends in US employment: 1860–1980. Source: Porat, 1977.

assumptions, since much depends on government policy. But some relevant points are worth listing:

(1) The fraction of the work force involved in direct (hands-on) manufactur- ing now is no more than 12% in the USA, and perhaps 20% in Japan. Social policy could easily reduce the size of the "potential" work force by reducing retirement age (probably undesirable) or increasing the num- ber of annual holidays and reducing the length of the workweek. Keeping people in school longer would also have this effect. A combination of the last two items alone, introduced gradually over the next 30 years, could easily soak up any manpower surplus with no "new" jobs created.

(2) Insofar as highly paid manufacturing workers constitute a significant el- ement of the middle class, some decline is almost inevitable. The im- portance of advanced education as a "ticket" to the middle class will increase, by contrast.

(3) A number of social and environmental functions are inadequately per- formed at present and could readily be expanded. Education, health care for the elderly and handicapped, and environmental restoration are only

three examples. The problem is how to pay for these tasks, which is basically an issue of tax and fiscal policy.

## Notes

[1] Life cycle length is not normally measured, and there are no comparable international statistics. However, according to surveys carried out in Japan, the average product life cycle for the entire machinery-manufacturing sector declined from 4.979 years in 1981 to 4.451 years in 1984, a drop of 10.6% in just three years (Mori, 1989, Table 1). In the electrical machinery subsector, the decline was from 4.553 years in 1981 to 3.246 years in 1984 (28.7%).

[2] For example, improvements in the precision of cylinder blocks, pistons, and piston-rings have already reduced the need for oil changes and extended expected automobile engine lifetimes. However there is still much room for improvement in quality and reliability of other automotive subsystems; such as power-assisted steering and braking.

[3] A negative of this sort is difficult to prove, of course, but we have not encountered any clear statement of the nature of the problem in the journals. However, it is interesting to note that, notwithstanding their well-known tendency to proceed incrementally (bottom-up), most Japanese firms have apparently recognized that in this respect CIM is "different." Surveys of Japanese managers taken in 1987 noted that 123 of 147 managers favored an incremental approach for the *present* but that only 32 felt it would be adequate for the *future* (Mori, 1989, Table 1). By contrast, only 14 of the 147 managers advocated a top-down approach for the present, but 103 felt it would be necessary in the future (ibid).

[4] It is not suggested that this "goal" is reachable, even in principle. On the other hand, it is extremely difficult to define a lower limit (other than zero) to the number of workers who will be needed to produce a unit of manufactured output in the distant future.

# Chapter 2

# Manufacturing Technology and Sources of Productivity

## 2.1 Sources of Past Gains in Manufacturing Productivity

The direction and pace of change in any technology can only be forecast on the basis of a solid grasp of the historical background. If the changes now apparent in the field of manufacturing technology are truly portents of a second (or third) industrial revolution, as I argue in this book, then it is not inappropriate to look back, at least briefly, at the changes that have taken place since the first industrial revolution, in the late eighteenth century.

The major innovation of the first industrial revolution (c. 1770–1830) was the substitution of steam power for water power and animal muscle power. This was of great importance in the UK, where good sites for water power were scarce to begin with and were essentially exhausted by the end of the eighteenth century. Horses, too, were expensive to maintain because of the high price of feed. However, in the USA, where animal feed was plentiful and water power was more readily available, steam power was introduced initially only for river and then for rail transport.

The immediate economic benefits of steam power (vs. water power), even in the UK, were quite modest – of the order of 0.25% per annum (p.a.) added to the annual growth of GNP – at least up to the 1830s when railroad building began in earnest (von Tunzelmann, 1978). Mechanization, the application of mechanical power (from water or steam) to drive textile machinery and woodworking or metalworking machines, seems to have been far more

**Table 2.1.** Mechanization vs. scale of production.

| Task category | Custom | Batch | Mass |
|---|---|---|---|
| Parts recognition and sorting | Manual | Manual | Not applicable |
| Parts transfer | Manual | Transitional (e.g., belt machine) | Mechanized (e.g., transfer machine) |
| Machine loading and unloading | Manual | Mostly manual | Mechanized (e.g., feeders) |
| Tool welding (including machine operation) | Semi-mechanized (manual control) | Mostly mechanized(NC) except for supervisors | Mechanized fixed sequence |
| Parts inspection | Manual | Manual | Transitional |
| Parts mating and assembly | Manual | Mostly manual | Transitional |

significant, in the long run. Mechanization and economies of scale made possible enormous increases in manufacturing productivity throughout the nineteenth century (*Table 2.1*, Chapter 1). However, the direct application of massive amounts of steam power to a single factory drive shaft peaked in around 1900, as electric drive began to be widely adopted, as shown in *Figure 2.1*.

Nevertheless, total installed horsepower per unit of output continued to grow at an average rate of 1.1% p.a. from 1899 until around 1920 (Schurr, 1984). It declined thereafter until 1953, and has increased slightly since then. Factory electrification (i.e., machine tools driven by electric motors) was highly beneficial in terms of flexibility of operations and plant layout. In fact, the adoption of electrified unit drive appears to be a major factor in the rapid improvement in US productivity growth that occurred after World War II (Schurr, 1984).

Yet, there were other major contributions to productivity gains since 1800. The most important historical milestone in the history of manufacturing, by many accounts, was the ability to produce truly *interchangeable parts* (e.g., Hounshell, 1984). This had been an explicit goal of military technology since 1717 (France). Interchangeability was achieved only by degrees.

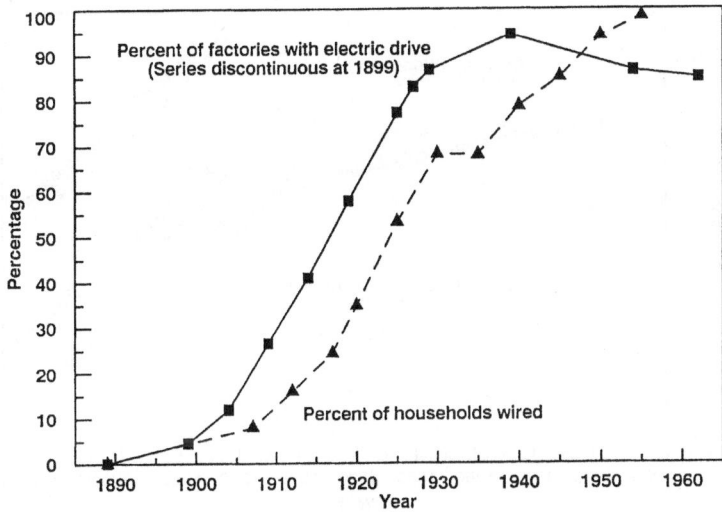

**Figure 2.1.** Electrification in the USA. Source: *Historical Statistics of the United States.*

A very crude form was claimed, for instance, by Eli Whitney (c. 1805), but it was not a practical reality even for weapons manufacturing until the late 1840s. Samuel Colt's famous exhibit at the Crystal Palace in London (1851) created a media sensation and undoubtedly marked a significant step in mechanization. It resulted in contracts for Colt to build munitions factories of his design for the British government.

Underlying the achievement of interchangeability was a series of innovations in precision, metalworking, and measurements by Wilkinson, Stowell, North, Whitney, Whitworth and Fitch, and others. The trend toward increased precision in measurement and in cutting has continued to the present, and even accelerated since World War II (*Figure 2.2*).

On the other hand, there is little or no evidence of major improvements in machine tool design since 1900. Modern production machine tools tend to be much bigger and more powerful than earlier counterparts, but they are scarcely more precise. Yet machine output per labor-hour input has increased enormously over the same time. For example, a 36″ vertical boring mill in 1950 operated by 1 worker could produce the same output in 1 day

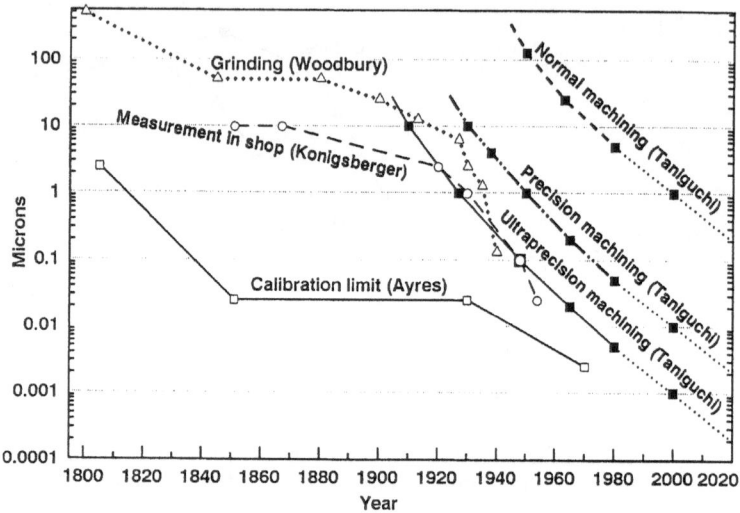

**Figure 2.2.**  Measures of precision in metalworking.

that would have required 50 such machines (and 39 operators) in 1890. Similarly, a 20″ engine lathe with 1 operator in 1950 produced the same output as 30 machines (and 50 operators) in 1890 (Tangerman, 1949). Similarly, the centennial issue of *American Machinist* (1977) cited a theoretical turned part – a steel axle – that would have required 105 minutes to machine in 1900, as compared with less than 1 minute in 1975.

Yet, based on machine tool attributes listed in catalogs, machine tool productivity – with characteristics held constant – should have *declined* more or less continuously at about 2% per year since the 1890s (Alexander and Mitchell, 1985). The most likely explanation of the Alexander–Mitchell paradox is that harder metals introduced since 1900 permit higher cutting speeds and less frequent tool changing. Prior to the mid-nineteenth century, the hardest available metal for cutting was carbon steel made by the crucible process (c. 1740) and "case-hardened" by heat treatment. A major step forward was the introduction in 1868–1882 of manganese-wolframite-based "self-hardening" alloys by Robert Mushet (Tylecote, 1976). These were the predecessors of "high-speed" tungsten steels developed especially by Frederick W. Taylor and Mansell White (c. 1900). This last development resulted in an approximately 70% increase in the maximum cutting rate from 1900 to 1915. The introduction of cemented tungsten carbide cutting tools resulted in cutting-speed increases of the same magnitude between 1915 and 1925.

Another major innovation was tungsten-titanium carbide, introduced by McKenna in 1938. Somewhat surprisingly, although few new cutting tool alloys have been introduced since then, tool fabrication (e.g., hardcoating) techniques have resulted in surprising further gains. Maximum cutting rates increased by no less than a factor of 10 from 1925 to 1975 (*Figure 2.3*). Interestingly, rapid improvements in cutting technology are still continuing, but the most recent gains are primarily due to advances in gas-bearing technology that will permit cutting speeds, in principle, at least 10 times greater than 3,000 sfpm (speed, feet per minute) achieved by off-the-shelf machine tools in 1977 (*American Machinist*, 1977). Machine tools have, once again, become a dynamic technology.

Continuing gains in cutting speed have not been matched by comparable improvements in other areas of manufacturing, unfortunately. In the early nineteenth century, manufacturing labor was predominantly concerned with wood or metal cutting and forming, but by 1900 progress in metalworking together with increased product complexity had changed the nature of the problem. The assembly of a complex product such as a clock, sewing machine, or bicycle – supposedly made from standardized interchangeable parts – typically constituted a labor-intensive activity requiring highly skilled "fitters." This was particularly true in Europe, where the greater availability of skilled labor resulted in a greater emphasis on high-quality (better finished) manufactured products as compared with the USA, where there was a greater emphasis on large-scale production at minimum cost.

By some accounts Henry Ford's historic contribution to *mass production* was achieved primarily by enforcing rigid quality control in parts manufacturing – utilizing the scientific management methods of Frederick W. Taylor (1911) – thus, finally eliminating the need for fitting. Ford himself stressed the combined principles of "power, accuracy, economy, system, continuity and speed." Ford engineers certainly looked everywhere for opportunities not only to subdivide the manufacturing process into many individual tasks, and to increase the efficiency of tasks by application of Taylor's methods, but also to substitute machines wherever possible for human workers. "Bringing the work to the man" was one of the ways to increase efficiency. Conveyor belts and gravity feeders began to be introduced extensively in the Highland

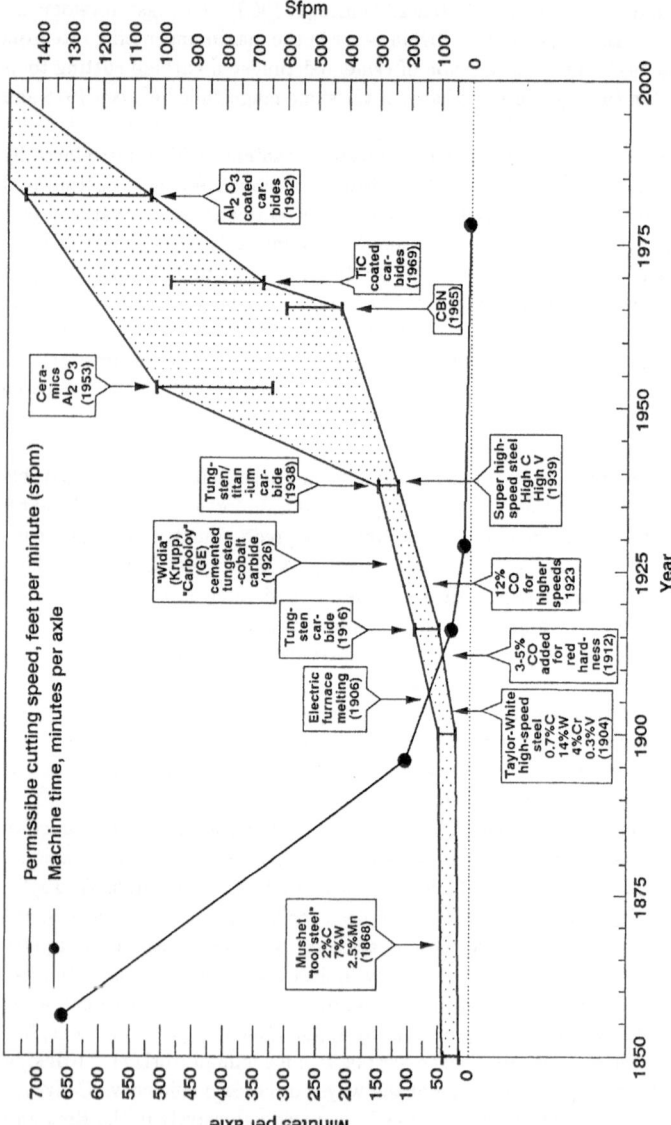

**Figure 2.3.** Machining speeds for steel axle: Machining time and permissible cutting speed.

Park plant by 1913. The moving assembly line (c. 1916) was the logical outcome of this rationalization.

Ford's assembly-line methods did, in fact, sharply reduce the cost of assembly as compared with parts manufacturing in the 1920s. However, in a fundamental sense, the assembly line is nothing more than a scheme to permit a more effective division of labor. The technology of assembly itself has changed very little until the last decade or so, except to the extent that assembly-line workers have gradually acquired power-assisted tools (such as wrenches) and the like.

In summary, while the mechanization of parts manufacturing has not yet reached any physical limits, its contributions to gains in manufacturing productivity were becoming negligible by the 1970s. In fact, logistics, assembly, and quality control now account for most of the direct costs of manufacturing – quite apart from *indirect* costs of R&D, engineering, finance, marketing, personnel management, and the like. To reduce costs significantly – and remain competitive – a completely new technology of production seems to be needed. This imperative will become increasingly manifest over the next several decades.

The historical factors resulting in productivity growth in manufacturing, from the first industrial revolution to the 1970s, can be summarized as follows:

- Division and specialization of labor.
- Application of mechanical power (from water or steam engines).
- Development of better engineering materials (iron and steel).
- New tools suitable for mechanization (turning, milling, and grinding).
- Methods of precision measurement (calipers, gauges, and comparators).
- Interchangeability of parts (elimination of "fitting").
- Electrification of machines to increase efficiency and flexibility.
- Harder materials for faster cutting tools (alloys and ceramics).
- "Scientific management" and vertical integration (Taylorism–Fordism).
- Mechanical integration (automatic transfer machines).
- Statistical quality control (SQC) and total quality control (TQC).

None of these factors involved computers in any fundamental way. Most had already reached a stage of maturity (or saturation) such that further gains would be expected to be very slow and costly, at best. The major exceptions – areas of rapid progress – were (and are): (1) the development of new engineering materials and corresponding new methods of fabrication (e.g., ceramics and composites), (2) greater precision, and (3) faster cutting speeds.

## 2.2   Manufacturing Operations for Non-Engineers

This is not the place for an extended discussion of manufacturing technology. Yet it is impossible to discuss changes in manufacturing technology, and still less the *implications* of those changes, without a minimal understanding of certain basic facts. Since the subject is seldom taught in an integrated way, it is fair to assume that few economists or managers who were not trained as mechanical engineers have any real idea *how* metal (or other) products are made. By the same token, few mechanical or electrical engineers have any clear idea of the manufacturing system as a system, or of the economic relationship involved. The first of these topics is reviewed briefly here; the second is discussed in Chapter 3.

### 2.2.1   General organization and layout

The work plan in any factory can be schematically represented as a hierarchy of basic *part manufacturing* (i.e., shaping and forming) *units* feeding parts to *subassembly stations* and thence to *final assembly*. The basic scheme is illustrated in *Figure 2.4*.

Suppose the "final" product (say an automobile) consists of $N$ different subassemblies, some of which are used singly while others are used in multiples. For instance, an automobile requires only one chassis and drive train, but four wheels and brake subassemblies, two headlights, four or six spark plugs, etc. (see *Figure 2.5*).

The $i$th subassembly, in turn, consists of $m_i$ different parts, some of which are used singly, while others are needed in integer numbers. Assuming, for the moment, that no parts are common to more than one subassembly, the final product requires $M$ different parts where

$$M = \sum_{i=1}^{N} m_i \ . \tag{2.1}$$

Each part must be produced by a distinct group of machines. In large volume production plants, these are typically linked together in either a rotary or linear sequence by an automatic workpiece *transfer system* of some sort. In lower volume facilities, parts may be transferred by belts, carts, or even by hand.

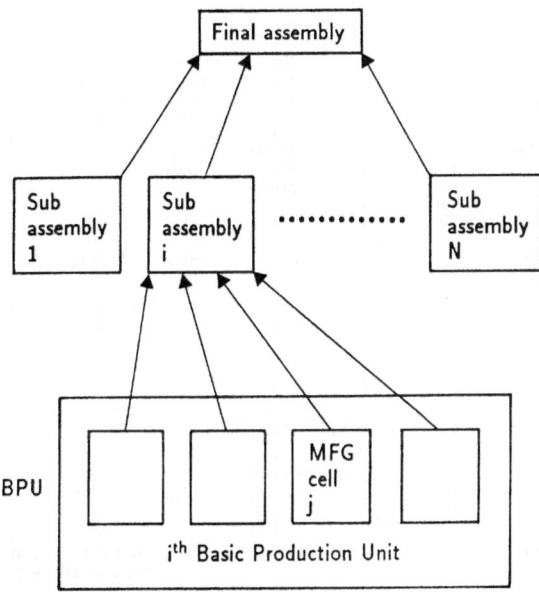

**Figure 2.4.** Factory organization.

In the extreme case of a cell dedicated to one variant of one part, the effective rates of all machines must be identical, since each machine does one and only one operation on each workpiece at a pre-specified rate before passing it to the next machine. If more than one variant of the part type is to be produced, a more flexible parts transfer system may be necessary. Such a group of machines, linked together and dedicated to making a single part, or variants of a single part, can be called a *flexible manufacturing cell* (FMC) or a *flexible machining system* (FMS). Machine layout in a cell depends on the degree of specialization. For a very specialized cell for a single part, where machines are linked by an indexing transfer line, the layout is likely to be linearly *sequential* (parallel) as shown in *Figure 2.6(a)*. For less specialized cells with more flexible transfer systems, machines might be grouped for various other sequences of operations, as in *Figure 2.6(b)* or *2.6(c)*.

For the $i$th subassembly, there will be $m_i$ different cells operating in parallel, producing parts that must be brought together at an assembly station.

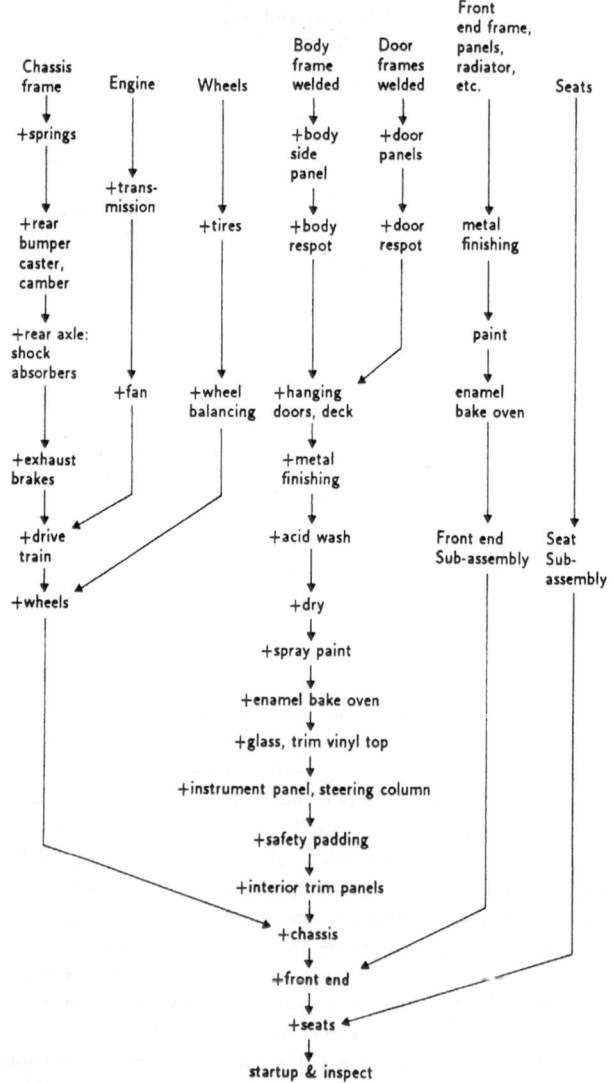

**Figure 2.5.** Auto assembly.

(a) Line layout

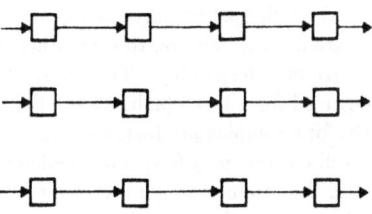

(b) Cell layout

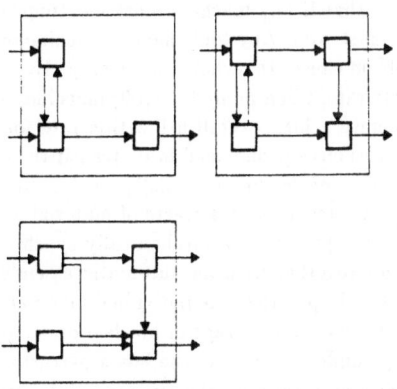

(c) Functional layout

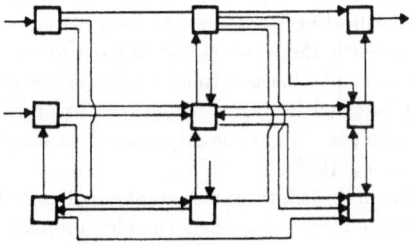

**Figure 2.6.** Alternative machine layouts.

Just as the effective rate of all machines in a given cell must be the same (for a given part), so all the cells feeding parts to a given assembly station must do so at rates that are fixed in relation to each other, based on the number of copies of each part in subassembly. The group of manufacturing cells producing parts destined for a given (sub) assembly station can be called a *cell-cluster*. Clearly, for a complex product, such as an auto engine or the car itself, a number of cell-clusters may feed subassemblies to a higher-order assembly plant. Again, the product design dictates relationships between the operating rates of the various clusters, which in turn, dictate the operating rates of the individual cells.

A key characteristic of the so-called *synchronous* sequential mechanical transfer system is that if any machine must be stopped for any reason, all of the machines must stop. Any tool change, machine breakdown, blockage, or jamming problem causes the whole line to stop. Even though such events are comparatively rare, when many (15–100) machines are linked serially, it is difficult to achieve a high overall utilization rate for the equipment. In addition, if any workpiece is damaged in its transit through the cell, it must either be removed – *unbalancing* the line[1] – or continue through the later steps in the sequence resulting in a waste of material and machine time.

The coordination problems of mechanically synchronizing a number of such transfer lines together with an automatic assembly system are obviously much greater. In practice, the individual manufacturing cells are usually *buffered* by intermediate storage of finished or semi-finished parts. The more such storage buffers there are, the less a perturbation at one location propagates disturbances through the system. But parts storage is costly, both in terms of capital tied up in incomplete workpieces and in the need for investment in specialized storage devices (usually top-loaded "towers" with a gravity-driven spiral track or chute). In some cases, pallets or magazines are needed for the workpieces to ride on while retaining their physical orientation, while facilitating mechanical loading of the machine at the next work station. Evidently there are significant costs to the use of intermediate storage in a factory. The "just-in-time" (JIT) system pioneered by Toyota and widely implemented in Japan has proved to be highly cost-effective, partly because it imposes higher-quality control standards on the feeder input streams (Hartley, 1986).

The discussion of the last three paragraphs applies to the case of a factory *dedicated* to a single model of a single complex product. The coordination problem is obviously much more complicated when the plant must produce a number of different models or variations of the product, or even a number of different products, with variable and uncertain production runs. If the designs are allowed to evolve over time, the coordination problem is still more complex and difficult. It is easy to see that dedicated machines linked by mechanically synchronous transfer systems are not applicable at all in such an environment. Each manufacturing cell must be somewhat flexible in terms of its input requirements, operating rates, and output specifications.

In the case of a dedicated line (or cell), each machine tool is specialized to a single operation, for which it can be optimized by design. Tool changing is thus minimized: tools are replaced at fixed intervals based on a precalculated useful life. Specialized jigs and fixtures (holding devices) are permanently installed on each machine, corresponding exactly to the specified shape of the workpiece and the specified requirements of the operation. Machines are designed to execute a fixed sequence of motions at a single optimum speed, to minimize the costs of each specific operation. On the other hand, in a "flexible" cell the individual machine tools must be capable of operating at various turning and cutting speeds, angles of attack, cutting depths, and so forth. Coordinating the machines in a flexible cell with the parts-handling system to achieve maximum output becomes a formidable challenge. Indeed, realistic cases are mathematically intractable: exact solutions cannot be computed. Such cases can only be analyzed by means of simulation and crude approximations (Wright and Englert, 1984).

In a traditional *job shop*, producing custom prototypes or small batches, the coordination of work flows to maximize productivity is carried out by the shop steward using information gathered from the individual machine operators plus his or her own accumulated experience. In Japan this technique is known as Seiban. In practice, however, the optimization problem is so complex that there is little attempt to do more than coordinate and avoid major bottlenecks. Usually, each machine is independently set up to carry out one (or more) operations on a certain number of workpieces based on the number of items in the batch. To be "on the safe side," the steward is likely to order extras of each part to accommodate mistakes or faults. The partially completed workpieces go into intermediate storage – most likely a bin – while the machine is set up again for something else. As the overall work load permits, the steward eventually assigns another machinist, on another machine, to set up and run the next operation or operations in the sequence. Fairly high machine and worker utilization can be achieved by completely separating each operation from the next in sequence. But the price of doing this is to stretch out in-plant transit time for each batch, which means carrying a large inventory of unfinished parts. In fact, in most job shops, workpieces are actually being worked on a very small fraction of the turnaround time between receipt of order and delivery to the customer.

In such a shop, incidentally, the machines are likely to be laid out by function (e.g., drill presses, punch presses, lathes, milling machines, etc.) as illustrated in *Figure 2.6(c)*. An important step forward in shop layout and scheduling was the classification of parts into families by shape.[2] Such classification systems are an important tool for deciding the way in which each part can best be produced (i.e., on what machine).

## 2.2.2   Shaping and forming operations

Most parts-manufacturing operations can be divided into two basic categories and some variants. The basic categories are *metal cutting* and *metal forming* to shape. The first involves *removing* excess material to arrive at the final shape, while the second involves little or no loss of material (in principle).

Variants of metal cutting include *turning* (on a lathe) and *boring*; *milling* (with a milling machine); *drilling, shaping*, and *planing* or *sawing*. All of these operations use a hardened, sharp-edged cutting tool, as illustrated in *Figure 2.7*. In addition, *grinding* should be regarded as a variant of cutting, inasmuch as it involves removing metal. However, the cutting is done by an abrasive, a very hard substance – usually a grinding wheel – made of alumina (emery) or silicon carbide (carborundum).

All metal-forming operations except rolling[3] use a *die* (or mold) to define the final shape. There are four subcategories, namely, *forging, shearing, casting* or *molding*, and *powder forming*. Final shape is achieved by applying pressure (or impact) or heat or both. Forging includes the following variants (*Figure 2.8*): *rolling, bending, punching, extrusion*, and *drawing*.

Shearing is a variant of pressing in which the metal is deformed beyond its breaking point, resulting in a cutting action. In practice, it is usually associated with punching. Casting and molding involve introducing the metal (or plastic) to the die in liquid form, where it hardens. Powder forming introduces the metal or ceramic material as a powder. It is subsequently compressed and hardened by sintering.

Despite the wide variety of possibilities, most metal-working operations involve cutting. The most common machine tools are milling machines, lathes (turning machines), and drills. Milling machines can generate virtually any surface shape. Lathes can generate any shape with external

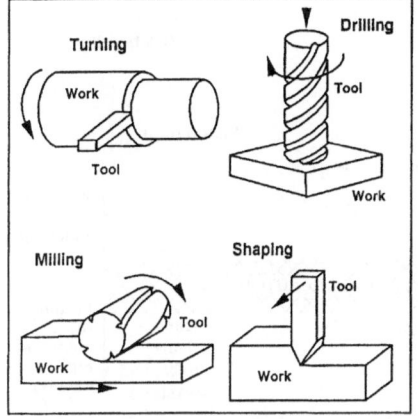

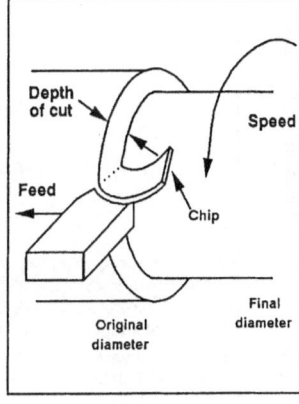

(a) The four basic machining processes can, between them, theoretically produce any contour on a work piece.

(b) In any of the basic machining processes, speed, and depth of cut determine productivity. The three variables are shown here for turning on a lathe

**Figure 2.7.** Basic operations in metal cutting. Source: Groover, 1983.

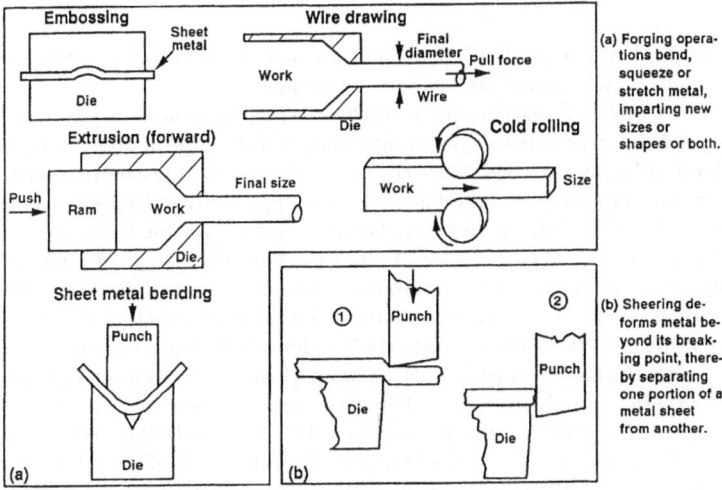

(a) Forging operations bend, squeeze or stretch metal, imparting new sizes or shapes or both.

(b) Sheering deforms metal beyond its breaking point, thereby separating one portion of a metal sheet from another.

**Figure 2.8.** Basic operations in metal forming. Source: Groover, 1983.

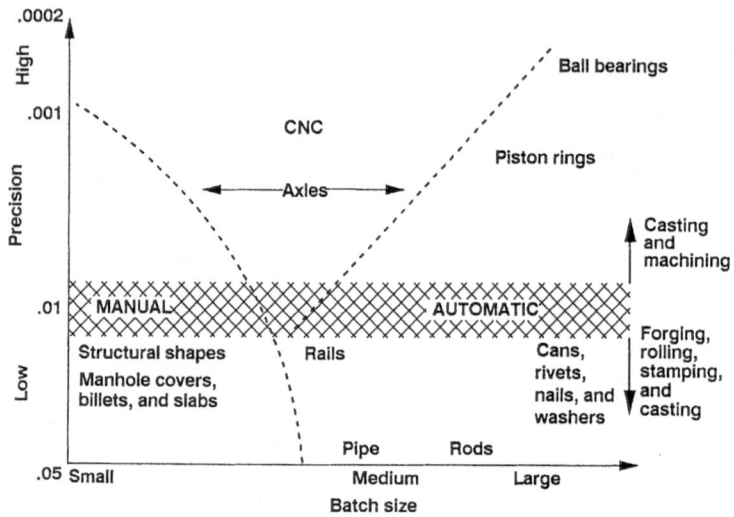

**Figure 2.9.**  Low-complexity shapes.

rotational symmetry. Drills are simply for making round holes. The most common metal-forming machines are punch presses.

The choice of machines (and machine types) to produce a given part depends mostly on its complexity, precision, and the scale of production, as shown in *Figures 2.9* and *2.10*. For low-precision simple geometric shapes, a forming process (casting, rolling, or stamping) is likely to be preferred because it involves little or no waste. Metal-cutting or grinding operations are required to achieve higher levels of precision. For small batch sizes, manual operations are used, whereas automatic machines are used for very large batch sizes. The first uses of computerized controls appeared in the small-to-medium batch region for relatively complex and high-precision parts.

Scale of production tends to be an inverse function of complexity (*Figure 2.11*) with relatively more parts of low complexity required than parts of high complexity. *Figure 2.12* illustrates a typical heuristic (rule of thumb) use in manufacturing, viz., that 20% of the parts constitute 80% of the value. (For obvious reasons, this is known as the "20–80 rule".) This relationship reflects the fact that parts with very simple geometries, such as washers, bushings,

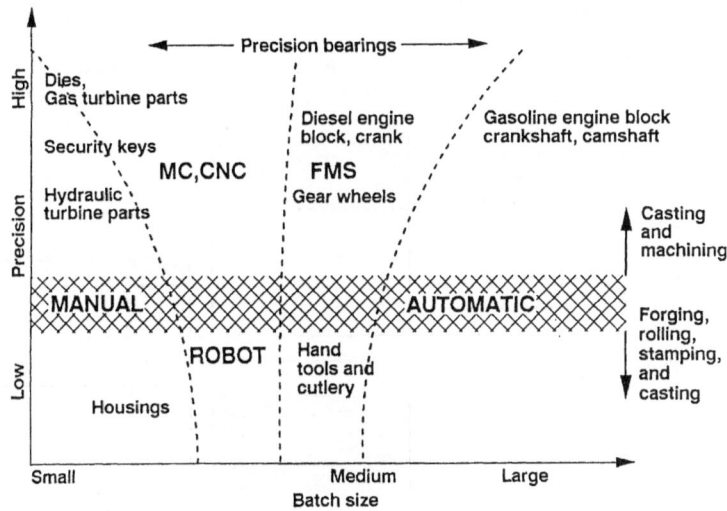

**Figure 2.10.** Very complex shapes.

or rivets, can be produced very inexpensively by specialized automatic machines (or "automats"), whereas complex parts require more elaborate and expensive machines. To assure that these machines are utilized efficiently, therefore, it is vital that they be *flexible* (i.e., *reprogrammable*).

The small, general-purpose, manually controlled tool in a typical job shop can be idle much of the time because the machine is comparatively cheap; the major cost is the skilled labor. To utilize high-cost skilled machinists efficiently, it is important that no machinist should ever have to wait for a machine to become available. But as individual machines have increased in power and precision, they have also risen sharply in cost. In a plant using expensive, multi-axis, high production rate tools the machines must be utilized a high percentage of the time. One way to achieve higher levels of machine usage is to reduce setup time by adopting NC or CNC.

The role of the skilled machinist is thus shifted gradually from that of machine operator to that of general supervision and setup. Once the program is prepared and calibrated, an NC machine tool can make complex shapes at a much higher rate than its manual-controlled predecessor. In fact, CNC

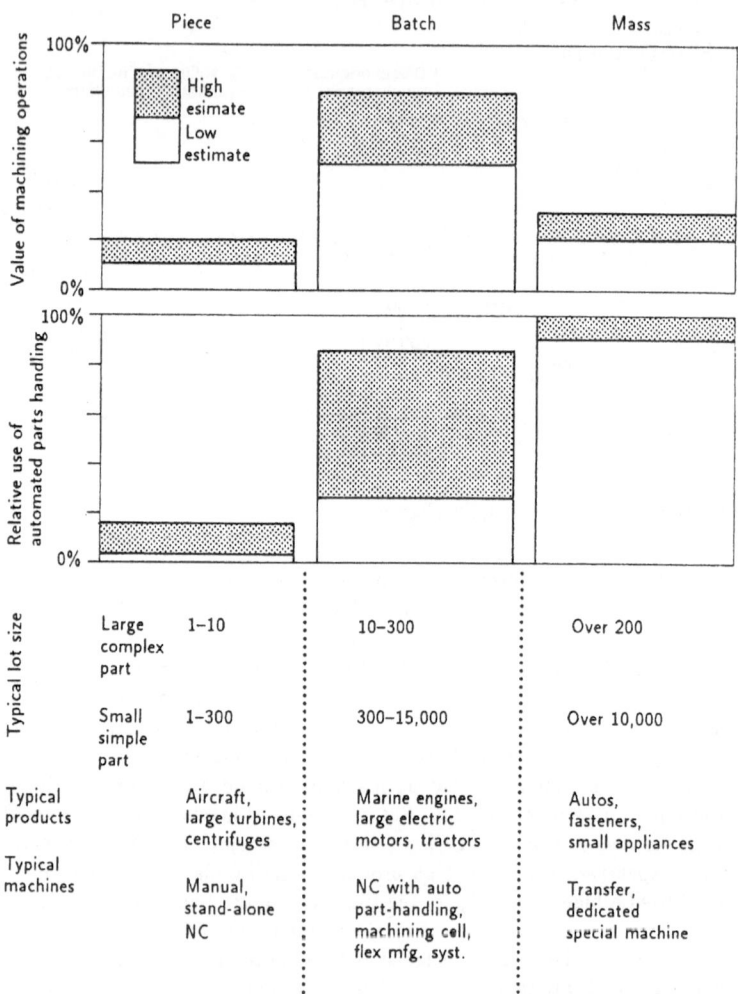

**Figure 2.11.** Production scale, variety, and complexity.

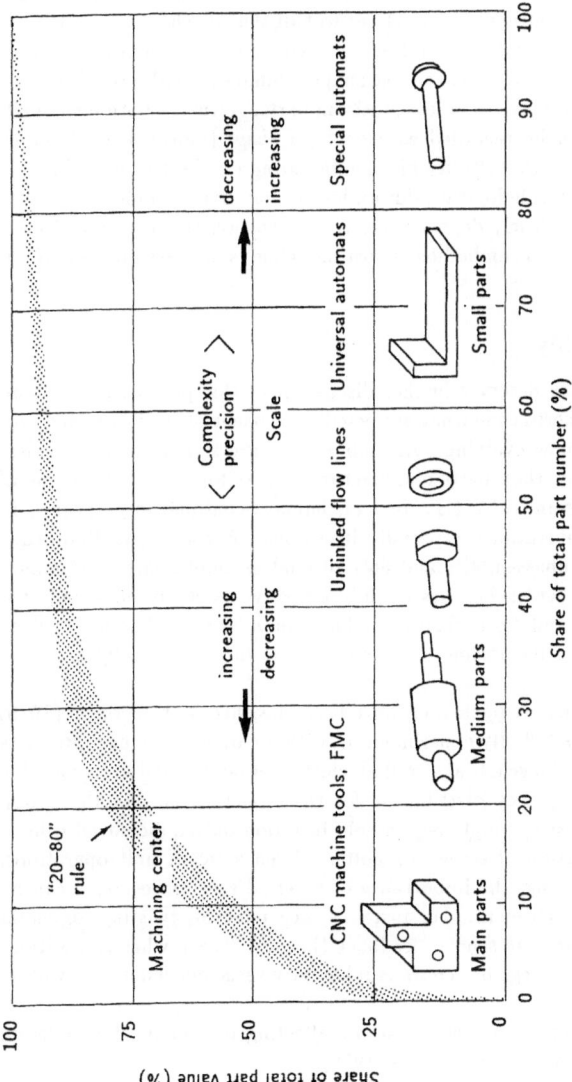

**Figure 2.12.** Part value vs. scale of production.

can increase output per machine by up to a factor of five. However, the average improvement is probably closer to half that much.

It is fairly evident that, for the same volume of production, complex shapes cost more to make than simple shapes. Similarly, high-precision parts are more costly to make than low-precision parts. Actually, both *complexity* and *precision* can be regarded as aspects of *shape* (morphological) *information*. Moreover, the activity of manufacturing can be thought of as the embodiment of morphological information in materials.[4] Indeed, the cost of manufacturing clearly depends, *ceteris paribus*, on the quantity of morphological information embodied. (Cost modeling is discussed in Chapter 3, Section 3.2.)

### 2.2.3   Assembly

The assembly task deserves further discussion at this point, since it is an essential and hitherto somewhat neglected component of the integrated production system now evolving. Assembly tasks are significantly less automated, at present, than parts-manufacturing tasks for the same volume of output, as shown in *Table 2.1*. In fact, humans continue to be needed in most assembly operations, especially insertions. According to Boothroyd (1980), 77% of subassemblies and 86% of final assemblies are done manually (c. 1980) either on benches or on progressive assembly lines; only 6% of subassemblies and 4% of final assemblies could be classed as mechanized (c. 1980). Alternative assembly systems are shown schematically in *Figure 2.13*.

The basic assembly operations have been classified by Kondoleon (1976) as shown in *Table 2.2*. By experimenting with a number of typical products (a toaster-oven, a bicycle brake, and an electric jigsaw), Kondoleon was able to ascertain a frequency distribution for these operations. For this group of products, at least, simple peg-in-hole insertion outnumbered all others, followed by insertion of screws or bolts. Given a set of unit operations, such as the foregoing, the importance of external sensory feedback can be determined by experiment. For instance, peg removal, twisting, or metal crimping do not require sensory feedback (in most cases), whereas positioning for insertion of pegs or screws is inherently sense-dependent, as will be seen later.

Some of the more important factors affecting positioning for insertion are as follows (Nevins and Whitney, 1978).

*   The amount of clearance (for free space) between parts after assembly.

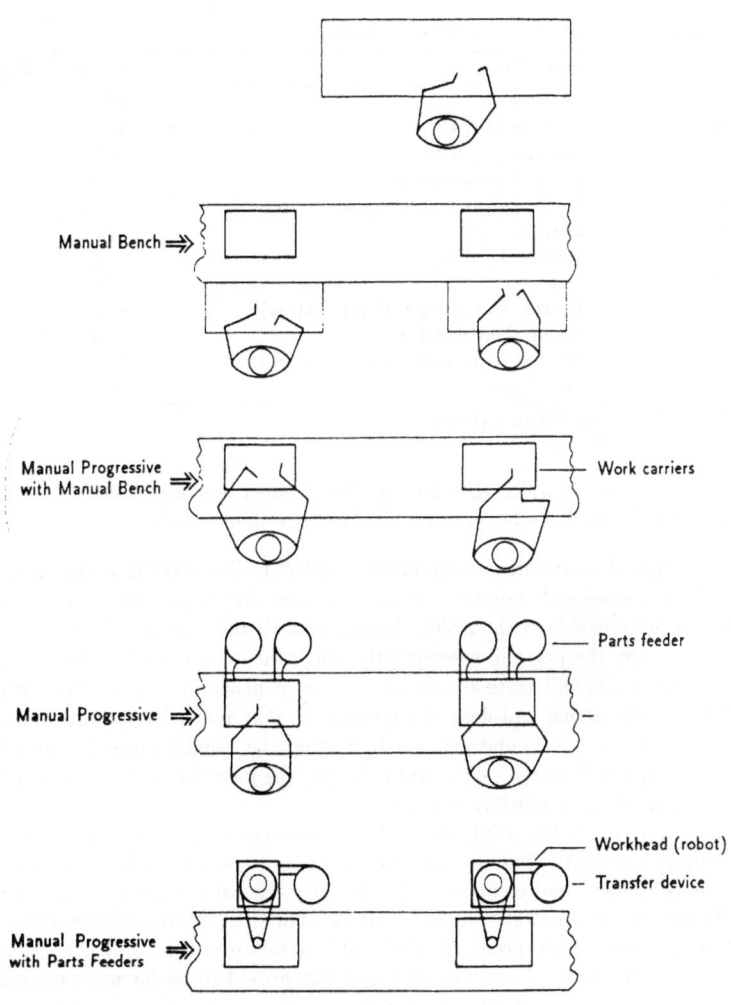

**Figure 2.13.** Alternative assembly methods.

**Table 2.2.** Typical assembly unit operations.

|   | Operation | Frequency (%) |
|---|-----------|---------------|
| A | Peg-in-hole | 34.5 |
| B | Push-and-twist | 12.8 |
| C | Multiple peg hole | 6.5 |
| D | Peg and retainer insert | 5.0 |
| E | Screw and/or bolt insertion | 26.8 |
| F | Force fit | 7.3 |
| G | Remove location pin | 1.0 |
| H | Flip over | 2.0 |
| I | Provide temporary support (fixture) | 1.5 |
| J | Crimp sheet metal | 0.5 |
| K | Remove temporary support (fixture) | 1.5 |
| L | Weld/solder | 0.5 |

Source: Nevins and Whitney (1978).


- The degree to which they are misaligned when they first touch.
- The friction force between parts when they slide together.

A typical positioning and insertion problem is illustrated in *Figure 2.14*. Holes are usually chamfered to aid in insertion. As the peg enters the hole, it touches one side of the inside channel first. If the angular misalignment is too large, the peg will subsequently touch the opposite side of the hole. Whether parts will mate successfully or not depends on the relative error between the actual and desired alignment as they touch for the first time. If the relative misalignment is small, mating can usually proceed without difficulty, but, if the misalignment is larger, the insertion will jam, causing an interruption or possibly damage.

To repeat one key point: the control function is inseparable from *sensory feedback* capability. This is true whether control is exercised by humans or "smart sensors" and computers. In the early days of computers, it seemed obvious that the digital computer, with its ability to perform complex calculations at ultra-high speeds, should be able to take over this control function quite easily. In fact, however, this goal has proved to be far more elusive than was originally believed. The problem – at least since the mid-1970s – is not a lack of raw computational power, nor is it the cost of computation *per se*. The problem lies with other elements of the control system, notably the sensors and the associated sensory-interpretation software and hardware. I discuss this topic further in the Section 2.3 and in Chapter 4.

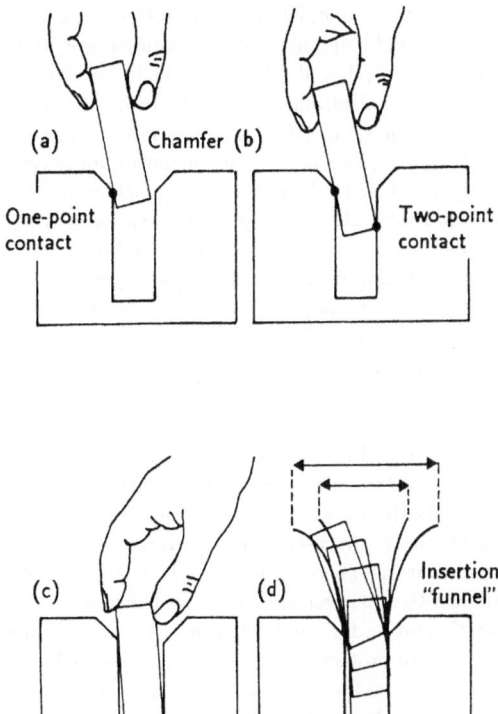

**Figure 2.14.** Tactile feedback in assembly.

## 2.3 Human Factors: Man–Machine Interface

In the economist's "bird's-eye" perspective, the manufacturing sector (as distinguished from mining, construction, or services) is devoted to the conversion of raw materials into finished and portable products ranging in size from tiny electrical components or fasteners up to ships and in complexity from nails to supercomputers. Basic activities can be divided into several categories: materials processing (refining, alloying, rolling, etc.); parts manufacturing (cutting, forming, joining, and finishing); parts assembly and packaging; inspection, and shipping, storage, maintenance, sales, etc.

Materials, energy, capital, and labor are said to be "factors of production." As a rough generalization, factors of production are regarded as substitutable for each other, i.e., labor or energy inputs can be decreased by increasing capital inputs. (This is not true, of course, for materials actually embodied in the product or for fuel to run the machines.) However, upon closer scrutiny, such substitutions are typically possible only at the margin and in a rather special aggregated sense. To clarify this point, consider the role of fixed (physical) capital, disregarding liquid working capital for the moment. Capital plant and equipment are of several distinct kinds: tools, dies, and patterns; machine tools and fixtures; materials-handling equipment, e.g., pallets, conveyor belts, transfer machines, pipes, pumps, forklifts, cranes, and vehicles; containers, e.g., shelves, bins, tanks, and drums; and structures and land.

Machine tools do substitute for workers insofar as they wield tools such as hammers, drills, punches, saws, milling cutters or grinding wheels, files, or cutting implements similar in function to hand tools as used by human workers. Machine tools are now used almost universally in manufacturing (at least, in developed countries) because they can be faster, stronger, and more accurate and tireless than human workers using hand tools. Motor vehicles are used for transportation for similar reasons. Containers and structures are required to store and protect materials in process as well as shelter tools, machines, and workers from the elements. Clearly, these categories of capital are somewhat complementary; at any rate, capital in one category cannot substitute for capital in another. Traditionally, the substitution of capital for labor has meant greater employment of machine tools in place of manual tools, and motorized forms of transportation in place of non-motorized ones. But, until recently, each machine has needed a human operator. In short, machines have been substituted mainly for human muscles but not for human senses and intelligence. In the past, machines and their human operators have been effective complements. The question implicit in the title of this section can now be made explicit: To what extent can machines be expected to take over other (control) functions of human workers in the near future, thus becoming direct substitutes for human labor?

To elucidate this question, a better functional taxonomy of repetitive factory tasks that are directly related to fabrication or assembly of parts is needed. (For present purposes, workers whose jobs are non-repetitive can be ignored, that is, personnel involved with building or machine maintenance, setup, scheduling, inventory, transportation, product design and testing, administration, or sales.) The major generic, i.e., repetitive, task categories are parts recognition, sorting, and selection; machine-parts transfer (loading and unloading); tool wielding; parts inspection; and parts mating, i.e., assembly.

All of these generic tasks can be accomplished, in principle, either by machines or by human workers. The most common patterns of automation in factories today are shown in *Table 2.1* and *Table 2.3*. In custom (or small-batch) manufacturing, most control tasks are and will remain largely manual, simply because it is not worthwhile to mechanize any task that is not highly repetitive. The increasing use of programmable machine tools in small shops does not contradict this conclusion. It merely reflects the fact that NC machine tools are becoming easier to program, so that microprocessors are able to control operations that can be entirely committed to memory, in advance. In larger-batch manufacturing, machine-tool loading and unloading is gradually being replaced by robots or programmable feeders, whereas assembly remains largely manual, though machine-assisted. Insensate robots also perform some tool-wielding operations, such a welding, spray painting, and gluing. In mass production, mechanization now extends to virtually all tasks except for magazine or pallet loading, inspection, and assembly, and even these are increasingly machine-assisted.

Although statistical evidence of skill levels over time is scarce, it is likely that the need for intelligence and skill in repetitive factory tasks on the factory floor has generally been decreasing for several decades. In effect, much of the skill formerly needed by a machinist, for instance, has now been embodied in sophisticated machines. This tendency was perhaps anticipated by many writers in the nineteenth century, but it was empirically confirmed for the first time by a four-year study conducted by James Bright (1958, 1966) of the Harvard Business School. Bright constructed a 17-level "automation ladder" (*Table 2.4*) and summarized the results of his observations and conclusions with respect to skill requirements, shown in *Figure 2.15*. It should be noted that factory automation today is, on average, several steps more advanced on Bright's ladder than it was in the late 1950s, with many applications exemplifying levels 12–14.

In virtually all cases, the remaining non-mechanized but repetitive factory jobs of today seem to require a significant level of tactile or visual sensory feedback. In fact, it is quite realistic to regard most factory workers in the semiskilled job classification as "operatives" (Bureau of Labor Statistics

**Table 2.3.** Comparison of manual manufacturing steps elimination by various degrees of automation.

| Step | | Conventional | Stand-alone NC | Machining center | FMS |
|------|--|--------------|----------------|------------------|-----|
| 1. | Move workpiece to machine | M | M | M | C |
| 2. | Load and affix workpiece on machine | M | M | M | C |
| 3. | Select and insert tool | M | M | C | C |
| 4. | Establish and set speeds | M | C | C | C |
| 5. | Control cutting | M | C | C | C |
| 6. | Sequence tools and motions | M | M | C | C |
| 7. | Unload part from machine | M | M | M | C |

M = manual operation; C = computer-controller operation.
Source: US GAO, 1976, p. 38.

terminology) or "machine controllers," to use a term that perhaps better conveys the essence of the human role in the production system.

In the modern manufacturing context, human factory workers can be modeled as part of an information processing feedback system. [This insight was expressed at least 35 years ago by Norbert Wiener (1948) in *Cybernetics*, and a number of early researchers in "human factors"/ergonomics.] They receive status information from the machine, the workpiece, and the environment. The workers process and interpret that information, arrive at certain conclusions, and translate the conclusions either into new control settings for the machine or into a new position/orientation for the workpiece. The amount of higher-order (problem-solving) intelligence required by the workers depends on how limited the set of possible responses is and how precisely the criteria for choosing among the responses can be pre-specified. In many cases, the workers need only decide whether the last operation was successful and signal for the next operation to begin. The major difference between jobs requiring semiskilled and skilled workers is that the former jobs involved relatively few and simple choices, each made many times, whereas

**Table 2.4.** Automation ladder: Levels of mechanization and their relationship to power and control sources.

| Initiating control | Type of machine response | | Power source | Level number | Level of mechanization |
|---|---|---|---|---|---|
| From a variable in the environment | Responds with action | Modifies own action over a wide range of variation | Mechanical | 17 | Anticipates action required and adjusts to provide it |
| | | | Mechanical | 16 | Corrects performance while operating |
| | | | Mechanical | 15 | Corrects performance after operating |
| | | Selects from a limited range of possible prefixed actions | Mechanical | 14 | Identifies and selects appropriate set of actions |
| | | | Mechanical | 13 | Segregates or rejects according to measurement |
| | | | Mechanical | 12 | Changes speed, position, direction according to measurement signal |
| | | Responds with signal | Mechanical | 11 | Records performance |
| | | | Mechanical | 10 | Signals preselected values of measurement (includes error detection) |
| | | | Mechanical | 9 | Measures characteristic of work |
| From a control mechanism that directs a predetermined action | Fixed within the machine | | Mechanical | 8 | Actuated by introduction of work piece or material |
| | | | Mechanical | 7 | Power-tool system, remote controlled |
| | | | Mechanical | 6 | Power tool, program control (sequence of fixed functions) |
| | | | Mechanical | 5 | Power tool, fixed cycle, one function |
| From a person | Variable | | Mechanical | 4 | Power tool, hand control |
| | | | Mechanical | 3 | Powered hand tool |
| | | | Manual | 2 | Hand tool |
| | | | Manual | 1 | Hand |

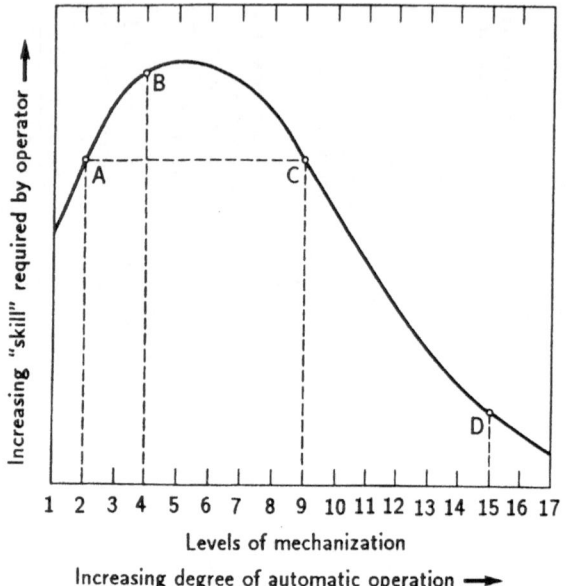

**Figure 2.15.** Labor productivity level (GDP per hour worked).

the latter jobs involve a very wide range of choices. Greater intelligence is involved when the range is so wide that each case is likely to be unique in some respect, requiring the worker to extrapolate or interpolate from known and understood situations; this is the essence of a non-repetitive job, of course.

Jobs in the goods-production sectors (agriculture, mining, manufacturing, and construction) are classified as *production* or *nonproduction*. The latter category includes office workers and sales personnel, those involved with logistics, etc. Within the "production" category, there is a further division between "direct" and "indirect." The former refers to workers whose labor is, in some sense, embodied in the product and whose wages are counted as *variable* costs of production. The latter refers to production workers whose labor is needed to keep the plant open and whose wages (or salaries) can be thought of as part of the *fixed* costs of production. (This distinction is somewhat arbitrary, to be sure. In Japan, it is reported that much of the direct labor is regarded as a fixed cost – with important consequences for long-term strategy. On the other hand, academic economists tend to regard all labor as a variable cost.) For the purpose of this book, indirect labor on the factory floor is largely concerned with engineering supervision and maintenance. It is essentially non-repetitive in nature. There are two different kinds of mental activity involved in doing direct manufacturing work, namely:

- *Process Monitoring*: decision making in response to external sensory data, regarding the state of the workpiece itself, or the state of the machines, tools, processing equipment, and/or the environment. Parts recognition and inspection are examples of "pure" monitoring tasks.
- *Motion Control*: decision making in response to either internal or external sensory data reflecting the physical state of the worker (but, in a more general ergonomic context, the notion of "workers" will later have to be broadened to include robots) in relation to the requirements of the task in hand. Most manipulative tasks, including assembly, involve motion to some degree.

These may be termed internal control decisions. The more efficient the process, the less control information, of either type, is necessary. The theoretical minimum amount of control information is that which is ultimately embodied in the product by that process. The rest of the information is ultimately lost.

Humans are able to process and reduce enormous amounts of information, relative to machines (including computers). The advantages of machines, as would be expected, are primarily on the output side, e.g., greater operating rate, power (or strength), and tolerance (or precision). It is helpful to consider these three variables separately.

(1) *Rate.* If weight and precision of location are not constraining factors, humans can identify and feed or transfer small parts, one by one, at rates of the order of one per second. Times for elementary motions have been compiled and published by the Maynard Foundation (Maynard *et al.*, 1948). Transfer machine magazine feeders and rotary bowl feeders can probably achieve consistently higher operating rates than humans for parts of a given size. However, the rate differences are small, perhaps factors of two or three, certainly less than a factor of ten.

(2) *Power.* Adult men in excellent physical conditions sustain a power output of the order of 250 W or more in short bursts, and 75–100 W for fairly long periods. (A world-class athlete, such as a swimmer or cyclist, may be able

**Table 2.5.** Human/machine performance ratios for generic factory tasks.

| Tasks/category | Measure | Human/machine performance ratio, $P$ |
|---|---|---|
| I    Parts transfer machine unloading | Rate | $10^{-1} < P < 1$ |
| II   Parts recognition and selection machine loading; parts machining; inspection | Rate/tolerance | $10^{-1} < P < 10$ |
| III  Tool wielding | Power/tolerance | $10^{-4} < P < 10^{-2}$ |

to generate 300 or more W of power output for several hours.) Machines, on the other hand, can be designed to deliver almost any amount of power. In practice, modern machine tools range in continuous effective power from 1 to 100 kW, or even more, depending on the application. Machines can outperform human workers in this regard by at least a factor of $10^2$ or $10^3$.

(3) *Tolerance.* Using hand tools and unaided eyes (or simple lenses), skilled human workers such as seamstresses, jewelers, and watchmakers can work to tolerances up to about $10^{-3}$ in. (or, perhaps, to $10^{-3}$ cm). Using mechanical and optical aids such as micrometers and microscopes, tolerances of $10^{-5}$ cm or better can be achieved by human workers such as engravers. Machine tools or automatic dimensional measuring devices with 1–3 degrees of freedom can be adjusted to move repetitively along paths or points in space with comparable precision. However, robots with more degrees of freedom tend to be about a factor of 10 less exact in repeating a motion than the most precise machine tools.

Taking into account these comparisons yields the cost-independent human/ machine performance ratios for all three groups of tasks shown in *Table 2.5*.

It is clear that the actual choice of a machine vs. a human for a particular task such as parts transfer is not determined, *a priori*, by a single comparison such as that in *Table 2.5*. If machines are expensive enough and human labor is cheap, of course human workers will be used. The comparisons in *Table 2.5* are more useful in providing insight as to the order in which various tasks are likely to be taken over by machines. This question is better addressed in terms of a two-dimensional comparison, as in *Table 2.6*, which takes into account all the human–machine differences noted above. From left to right, tasks are ranked roughly in terms of increasing intrinsic difficulty for humans, and from top to bottom in terms of increasing difficulty for machines. Evidently, tasks in the upper-left quadrant are hard for both.

**Table 2.6.** Intrinsic task difficulty: Humans vs. machines.

Increasingly difficult tasks for humans →

← Increasingly difficult tasks for machines

| | | |
|---|---|---|
| Pick and place a medium-sized oriented metal part<br>Spot-weld a repetitive pattern<br>Sandblast a wall<br>Spray-paint a simple surface<br>Drive a train on tracks<br>Arc-weld along a seam | Assemble a wooden cabinet, electric motor, or pump, repetitively<br>Pick and place a heavy metal part<br>Solder very tiny wire connectors<br>Spray-paint a complex surface | Pick and place a very heavy metal part in a hot, noisy (toxic) environment<br>Operate a fire extinguisher inside a burning building<br>Laser brain surgery (ex diagnosis)<br>Land a spacecraft, good weather |
| Pick and place a randomly oriented part (medium size) from a bin<br>Pick unripe fruits/vegetables by size only<br>Wash windows, selectively<br>Wash dishes and glassware individually<br>Inspect eggs in a hatchery | Inspect a printed circuit board for faults<br>Build a brick wall (3-D)<br>Cut coal from a face<br>Operate a farm tractor<br>Assemble a wire harness<br>Finish (e.g., lacquer) a cabinet, to order<br>Land a small plane (day, good weather, no traffic)<br>Cut and assemble a suit<br>Identify counterfeit paper money | Assemble a mechanical watch<br>Control air traffic at a busy airport<br>Weld a broken waterpipe from inside<br>Inspect a VLSI chip for faults |
| Pick and place very floppy objects<br>Cut and arrange flowers<br>Harvest ripe soft fruits/vegetables by color or texture<br>Separate crabmeat from shells<br>Inspect seedlings in a nursery<br>Plant seedlings | Drive a truck or bus through traffic<br>Deliver a (normal) baby and inspect for faults<br>Dental hygiene<br>Repair lace<br>"Invisible mend" a garment | Land an airliner at night, bad weather<br>Identify a counterfeit "old master"<br>Repair a damaged "old master"<br>High speed auto chase through city traffic<br>Diagnose a medical condition<br>Heart or liver transplant |

The first tasks likely to be automated are therefore in the upper right, the boundary being determined by the cost of labor vs. the cost of capital.

## 2.4  Manufacturing Productivity Trends

At the aggregate level, advances in technology are reflected in manufacturing productivity. Of course, manufacturing productivity (defined in terms of output per unit factor input) has been increasing more or less continuously since the beginning of the industrial revolution. Recent trends in labor productivity for the major Western countries at the level of economy as a whole are shown in *Figure 2.15*. Comparable data for manufacturing productivity broken down by a factor (labor and capital) are only available for a few countries. *Figure 2.16* compares labor and capital stock productivity for 1966, 1976, and 1986 for the manufacturing sectors of the USA, the UK, and the Federal Republic of Germany. It is noteworthy that while labor productivity has consistently increased, capital productivity has not. (In fact, it has generally decreased.) The productivity data for the USA is available in more detail, as shown in *Figure 2.17* and *Figure 2.18*, with the same pattern.

In the earliest period, great gains in labor productivity were achieved simply by specialization of tasks or "division of labor," as exemplified by Adam Smith's famous pin-making factory. A century later Frederick Taylor and others formalized and systematized the notion of reducing a complex job to a series of simple tasks that could, in turn, be scientifically analyzed and optimized. Further gains resulted from the substitution of water power or steam engines for animal or human muscles to drive machines. This permitted substantial increases in machine size and speed. Machine tool capabilities, in terms of "degrees or freedom" and accuracy, increased dramatically during the nineteenth century, culminating with the development of production-type grinding machines in the first decade of this century.

These changes were both facilitated and necessitated by the introduction of harder and stronger engineering materials, initially the substitution of steel for wrought or cast iron, and later the introduction of many specialized alloys (e.g., "high-speed" steel) and synthetic carbides for cutting and grinding. The last development, alone, is responsible for gains in machining speed of the order of 600-fold since the 1860s (*Figure 2.3*). Further gains in manufacturing productivity resulted from the substitution of individual electric drive for centralized shaft power in the first third of this century. Finally,

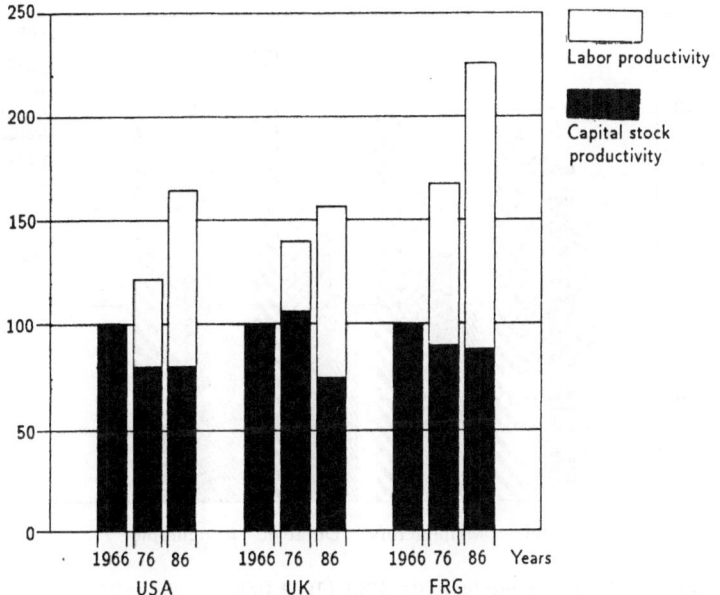

**Figure 2.16.** Labor and capital productivity, the USA, the UK, and the Federal Republic of Germany.

the moving assembly line and the mechanical integration of a large number of machine tools linked by a transfer line yielded further gains, culminating with the large automobile engine plants built in the 1950s and 1960s.

However, this last development seems to have signified the end of an era. While technological progress in conventional production technologies such as those noted above has not ceased by any means, the economic gains to be had from further increments in machine size, speed, or accuracy seem to be less and less significant. Two factors seem to be responsible. One is the increased competition in US and world markets resulting from the rise of the Japanese and other East Asian export-oriented economies. This has destroyed the postwar hegemony of General Motors in the (world) auto industry. It has also made obsolete GM's formerly dominant strategy of gradual "managed" innovation (the annual model change), with its emphasis

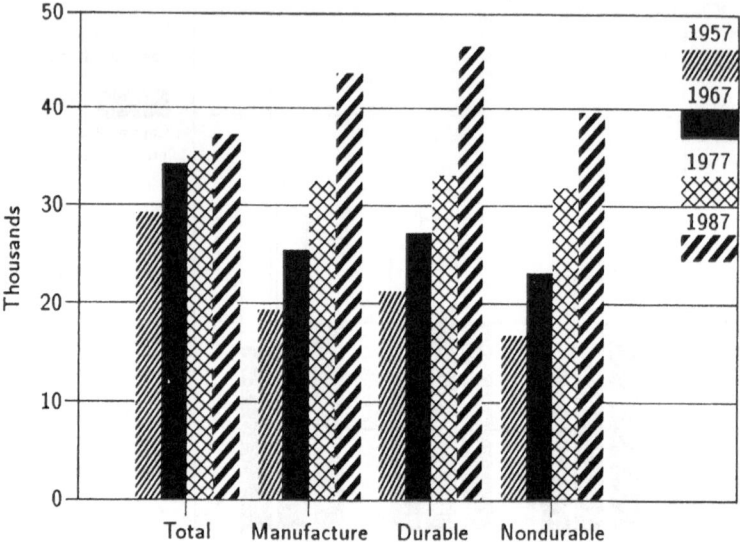

**Figure 2.17.** Labor productivity, USA (1982 US$).

primarily on exterior appearance. In current market conditions the relative importance of performance and quality vs. style has increased sharply, requiring a correspondingly greater emphasis on manufacturing. The second, and related, trend is toward increased product complexity, not only in the auto industry, but throughout manufacturing industry.

The consequences of the two trends are also twofold. In the first place, the rate of product design change has accelerated, in the auto industry and elsewhere. Whereas in the 1950s and 1960s the auto industry phased in design changes rather slowly so as to permit mass-production facilities a 20-year useful life before major renovation and retooling, today this is no longer possible. But, on the other hand, increased complexity has made the design process increasingly expensive and risky. This led to the so-called productivity dilemma: an apparent contradiction between the need to cut manufacturing costs by maximizing standardization and specialization and the need to introduce new and improved products (Abernathy, 1978).

The possibility of a way out of this dilemma through increasing the flexibility (economies of scope) of manufacturing technology was only clearly

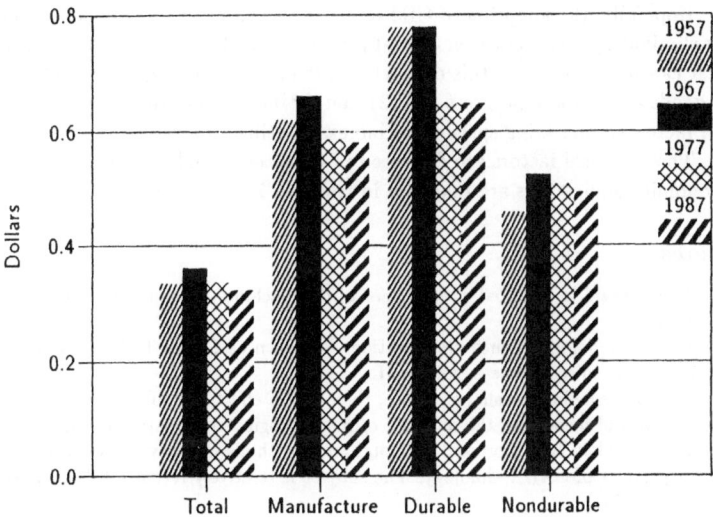

**Figure 2.18.** Capital stock productivity, USA (1982 US$).

recognized in the 1990s, although the facilitating technologies – computers and numerically controlled machines – began filtering into the manufacturing world about 30 years ago. I return to the issue of flexibility later.

In summary, it must be emphasized that labor productivity gains have been achieved in the past largely by increasing the magnitude of capital investment. This is the primary reason for static or declining levels of capital productivity. However, as the capital *intensity* of the economic system as a whole increases, capital costs account for an increasing fraction of total costs. Indeed, if savings rates remain constant and capital output ratios continue to decline (or even remain constant), a time must inevitably come when the replacement of depreciated capital consumes all of the surplus available for investment in the economy. At that point, economic growth would come to a halt.

Indeed, the worldwide productivity slowdown that has occurred since the 1970s strongly suggests that this might be happening, especially in the USA. Are we doomed to a no-growth economy for the indefinite future?

Hopefully, the adoption of CIM offers the way out of this dilemma, too, by eliminating constraints on capital productivity. The specific mechanisms that can be identified in this connection are (1) sharply reduced inventories of goods and work in progress and (2) sharply increased output per machine, via increased operating speeds and increased utilization rates.

The technical factors are discussed in Chapters 4 and 5, and the macroeconomic implications are taken up in Chapter 7.

## Notes

[1] A line is said to be unbalanced if some of the machines are working while others are idle.

[2] The methodology of such classifications is known as Group Technology (GT). For further discussion see Volume II.

[3] The rollers define the shape. Hence, they are, in effect, the die.

[4] Methods of measuring the quantity of morphological information in shapes are not yet widely known or used, although such methods will inevitably evolve as a by-product of CAD technology. The subject is relatively recent (Wilson, 1981; Ayres, 1987).

# Chapter 3

# Simple Economics of Manufacturing

## 3.1 Economies of Scale and Learning

The most salient feature of manufacturing, from the economist's point of view, is the pervasiveness of *economies of scale*. For a typical manufactured metal product that costs $1 per unit when it is mass produced in quantities of the order of more than 100,000 per year, the cost for a single copy made by traditional job shop methods would probably be more than $200. Or, take an example that I will return to later: an automobile engine of standard design costs about $2,500 prorated as part of the price of a $10,000 car. It would cost around $500,000 if all the parts were custom-made and assembled in a typical machine shop.

Economies of scale exist primarily as a consequence of the fact that some costs of production must be paid whether or not anything is actually made, but remain constant for wide ranges of output. Examples include capital charges, R&D, design, manufacturing engineering, and other overhead functions. These are called *fixed costs*. As output increases from zero toward the level of capacity, the fixed costs are spread over more and more units; thus, fixed cost per unit declines (roughly) inversely with increasing output up to the physical limits imposed by plant and equipment.

By contrast, there is another category of costs, called *variable costs*, that is more or less proportional to actual output. Examples include direct ("hands-on") labor, purchased materials that are embodied in the product, and utility services (e.g., electricity) used in the production process. In fact,

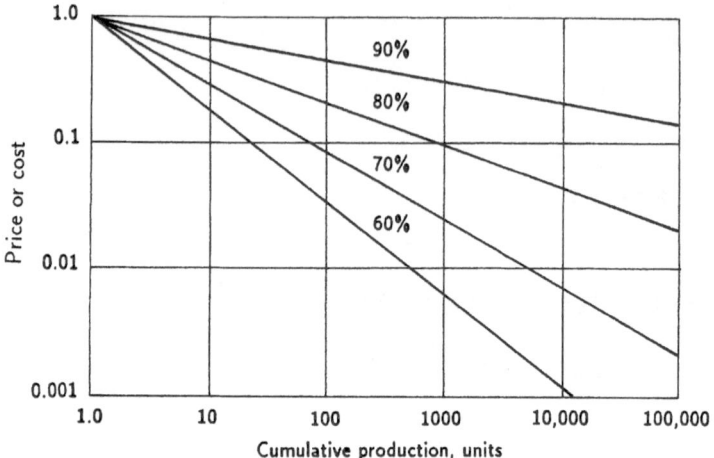

**Figure 3.1.** Standard form of learning or experience curve.

variable costs per unit do tend to decrease with scale of production, though not as dramatically as fixed costs per unit. In the case of labor, there is a well-documented tendency of unit labor requirements to decline as a function of cumulative experience. (A typical set of *experience curves* is shown in *Figure 3.1.*) For instance, the expression $a = .9$ means that each doubling of the total number $N$ of units produced ("cumulative experience") brings variable costs down to 90% of the previous level. One implication of the phenomenon is that firms with larger production levels accumulate experience faster. Thus, labor costs tend to decline, in general, with increasing scale.

For purchased materials, there are two elements of cost that depend on scale. In the first place, larger producers can justify more capital-intensive processes, which means that they can justify more expenditure on secondary recovery and reuse of waste materials; the result is that they use less net material inputs per unit of output. In the second place, larger producers can typically get better unit prices on purchased materials from their own suppliers, because larger orders can be processed more cheaply owing to the economies of scale enjoyed by the supplier! The result of these two factors is that variable costs decline significantly with increasing scale of output, up to the capacity of the plant.

A point of considerable importance, however, is that variable costs per unit tend to increase again, at some point beyond the *capacity limit*. In fact it is taken for granted, in standard neoclassical microeconomic theory, that marginal total costs begin to increase after total volume reaches some critical point determined by the plant configuration, equipment characteristics, and process. (If this were not so, it would never be necessary to build a new plant, no matter how great the demand. It would always be cheaper to increase the throughput for existing plants.) The fundamental reasons for increasing returns to scale beyond the capacity limit are not considered deeply by economists. However, ergonomists would tend to place the blame on some combination of bottlenecks and overload/interference phenomena (see, for instance, Ayres, 1987).

Since for every plant (and the technology embodied in it) there is a point beyond which unit costs start to increase, it follows that there is an optimum choice of technology for each level of output. In other words, there is a least-cost choice of production technology (for a given product-mix), which is also a function of scale. Reverting to the original example, if one were to manufacture "customized" automobile engines a few at a time, the method of choice would be to utilize a technology with the minimum fixed cost, regardless of variable cost. This would be a machine shop (job shop) with a number of inexpensive general-purpose machine tools and a staff of highly skilled machinists capable of doing almost any task. The machinists would be working steadily, moving from machine to machine. Each machinist in such a shop spends most of his or her time setting up the machines. This means reading the blueprints, deciding on the optimum sequence of operations, adjusting the clamps that hold the parts, and adjusting the machine controls to achieve the desired cuts. Both the workpieces and the machine tools themselves are idle most of the time. The allocation of time on the machines in such a facility in the USA (1980) might look like *Figure 3.2(a)*.

At an intermediate level of output (i.e., a medium-to-large batch of identical items), the sequence of operations would only have to be decided once. Moreover, various specialized devices would be introduced to increase the productivity of the machinists. For example, specialized jigs and fixtures would be made to order to hold the workpieces in fixed positions for particular machining tasks that have to be done many times. Parts feeders and rotary tables might be utilized to speed up parts handling. To avoid setting up the same job over and over, the more complicated setups could be automated by introducing numerical controls. Making use of geometrical similarities to different parts categories (known as "group technology"),

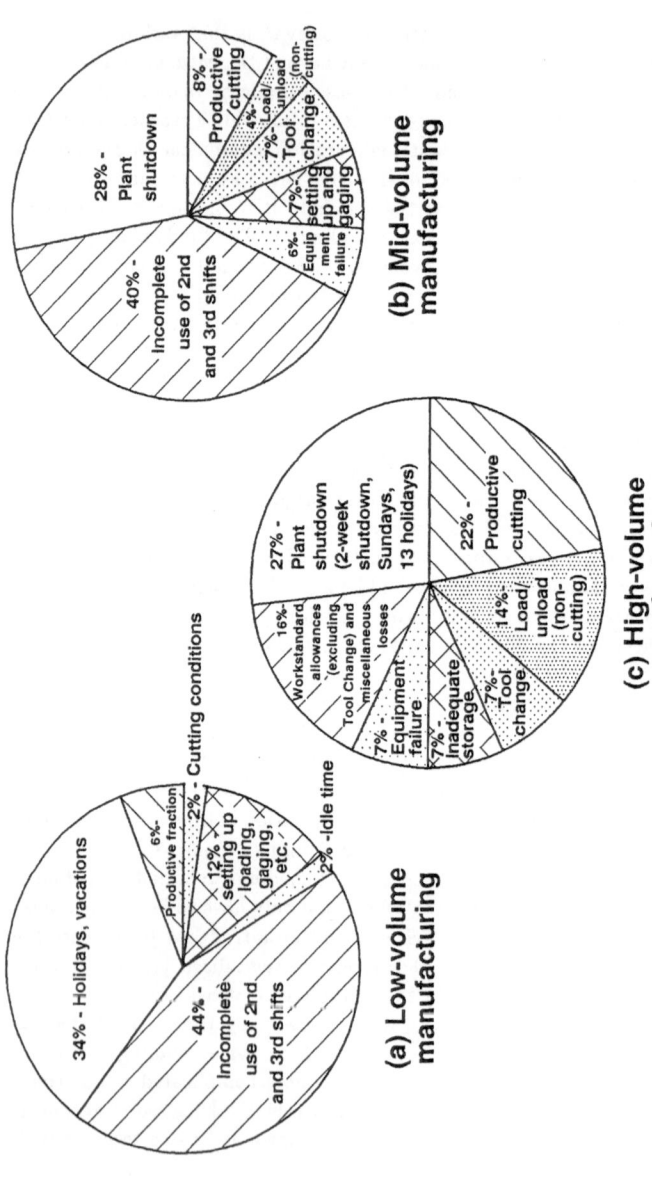

**Figure 3.2.** (a) Low-volume manufacturing; (b) mid-volume manufacturing; (c) high-volume manufacturing. Adapted from: *American Machinist*, 1980.

the machines could be physically grouped or clustered in such a way as to minimize materials transfer times.

For larger batch production such a group might also be connected by some sort of automatic transfer system and controlled by a minicomputer. This arrangement would be a manufacturing cell or a flexible manufacturing system (FMS), depending on its complexity. For medium-sized batch production of rather complex parts, it might also pay to buy a machine that could do many different successive machining tasks on the same piece, using automatic tool changers. Or it might pay to buy a specialized machine for doing one particular task, such as heat-treating. The overall effect of these changes would be to reduce the amount of time needed for setting up jobs (the biggest cost of skilled labor), and thus to keep the machines themselves busier with production, as shown in *Figure 3.2(b)*.

At very high levels of output of a single item, the optimum way to produce auto engines or anything else is to minimize variable costs (i.e., labor) as much as possible. The classic way to achieve this (c. 1975) is to design each machine to do one machining task only, and to connect the various machines together in a sequence by means of an automatic synchronous transfer line. This effectively makes the production line into a single machine, which carries the workpieces from one work station to the next, without ever being touched by human hands. For instance, the heart of an automotive engine plant would consist of a set of giant multiple-spindle machines with several hundred cutting tools (mainly drills and milling cutters) cutting simultaneously. Each spindle at each station is positioned very precisely with respect to all the others, and all are mechanically synchronized by means of elaborate gear-trains so that holes are precisely parallel and drilled to precisely the same depth.

This is the technology pioneered by Ford in the 1950s, and known as "Detroit automation." It still describes the situation in most high-volume automotive engine plants today. The only need for direct human labor is for overall supervision and for some of the trickier assembly tasks and inspection. Thus, the (roughly) 600 distinct machining operations required for a V-8 engine block involve only 1 minute of direct labor time, in such a plant, and cost about $25. The same operations in a job shop would take at least 600 minutes (10 hours) of productive labor time (Cook, 1975; Cross, 1982).

If every machine operated perfectly, without any breakdowns, then automated plant could operate continuously, 24 hours a day. In practice, however, breakdowns of various kinds are a fact of life, and the plant must be organized to cope with them. When a machine does stop for some reason,

all the machines in the line to which it is linked must also stop until the
problem is corrected. Experience suggests that a machine utilization rate
of 40% to 50% is about as high as can be achieved in practice, as shown in
*Figure 3.2(c)*.

## 3.2   A Simplified Cost Model

Taking into account the points discussed above, the cost curves for small-
batch (manual), medium-batch and mass-production technologies are shown
in *Figure 3.3*. Combining them into a single function of scale of production,
the typical long-run cost curve for metal products looks something like *Figure
3.4*.

In the upper left-hand corner of *Figure 3.4*, where production is customized ("one-of-a-kind"), fixed costs dominate. At the other end of the
spectrum, the lower right, variable costs are dominant and fixed costs per
unit are negligible. In between, the unit cost function, $C(N)$ has a form
something like

$$C(N) = V(N) + F/N \ , \tag{3.1}$$

where $N$ is the number of units over which the fixed costs, $F$, are spread and
$V(N)$ reflects the decline of variable costs, $V$, with increasing experience. A
typical empirical form for $V(N)$, would be

$$V(N) = AN^{-b} \ , \tag{3.2}$$

where

$$-b = \ln a / \ln 2 \tag{3.3}$$

and the parameter $a$ can be interpreted in terms of the different curves of
*Figure 3.1*. A number of historical examples are plotted in *Figure 3.5*. (It
must be emphasized that this simple formula is only intended to be a very
rough representation of the real world of manufacturing. Real cost models
are likely to be far more complicated.)

Apart from the magnitude of total costs, the *structure* of costs varies con
siderably from small-batch production to mass production. Cost allocations
on three different scales are shown in *Table 3.1*. It is important to realize
that the numbers in the table are rough averages, with wide variations from
sector to sector (see also *Table 3.2*).

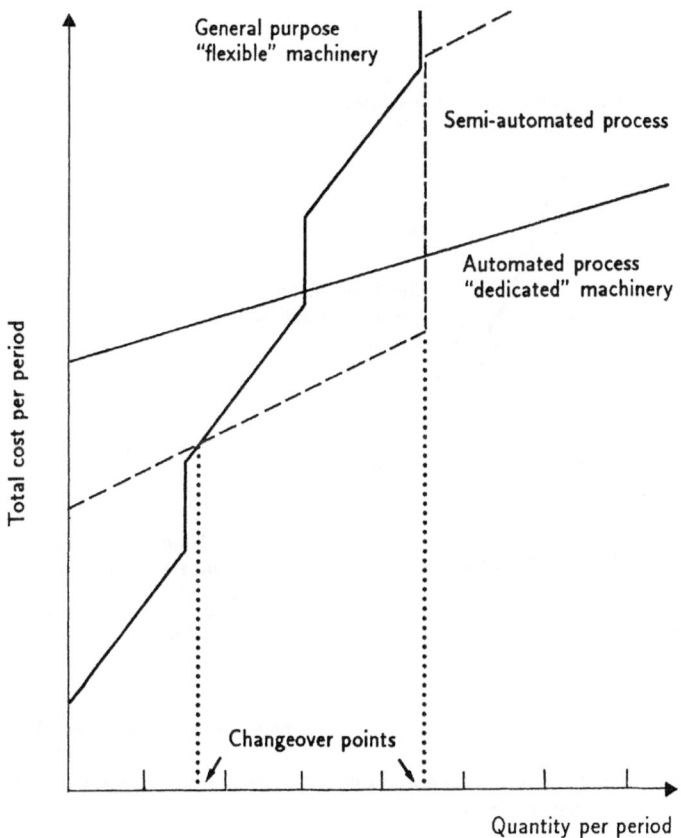

**Figure 3.3.** Cost curves for small, medium, and large batches.

The cost model described above deals explicitly with only one key variable, the number of units (scale), $N$, to be manufactured. The discussion in Chapter 2, especially Section 2.2, makes it clear that manufacturing costs are also dependent on the shape complexity (and precision) of the product. Both shape complexity and precision are reflected by a single variable: the quantity of morphological information ($H_{morph}$) required to specify the required shape to the required precision.[1] For a given scale $N$, both $V$ and $F$

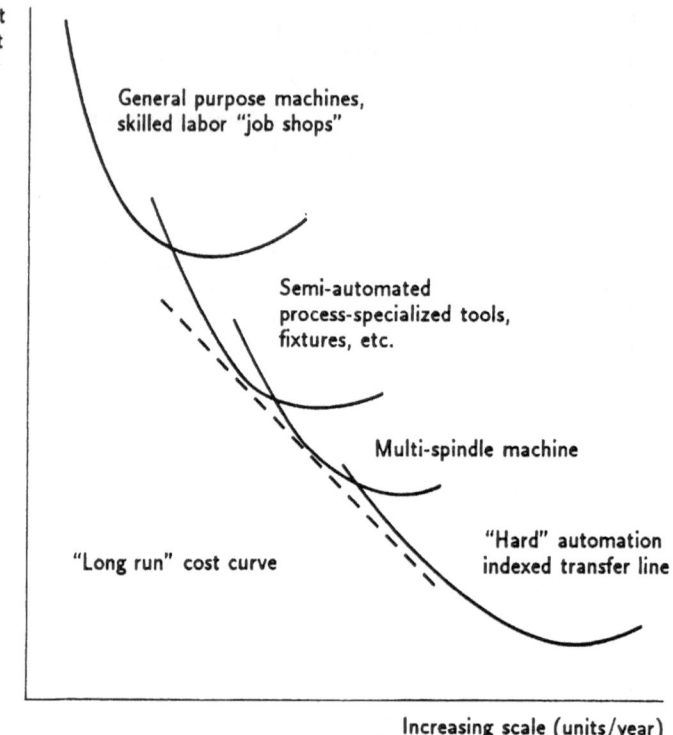

**Figure 3.4.** Long-run cost curve. Automation "embodies" skills and knowledge in systems and machines.

should also be some increasing function of $H_{morph}$. The best mathematical choice of this function can only be determined by future research. It should be noted that for a given complexity/precision ($H_{morph}$), $V$, and $F$ are also functions of the material from which the product is to be made. This reflects the well-known fact that some materials are easier to cut and form than others. Generally speaking, hard ceramics are the most difficult materials to cut and shape, followed by superalloys, steel alloys, carbon steel, cast iron, wrought iron, aluminum, zinc, brass, and various plastics.

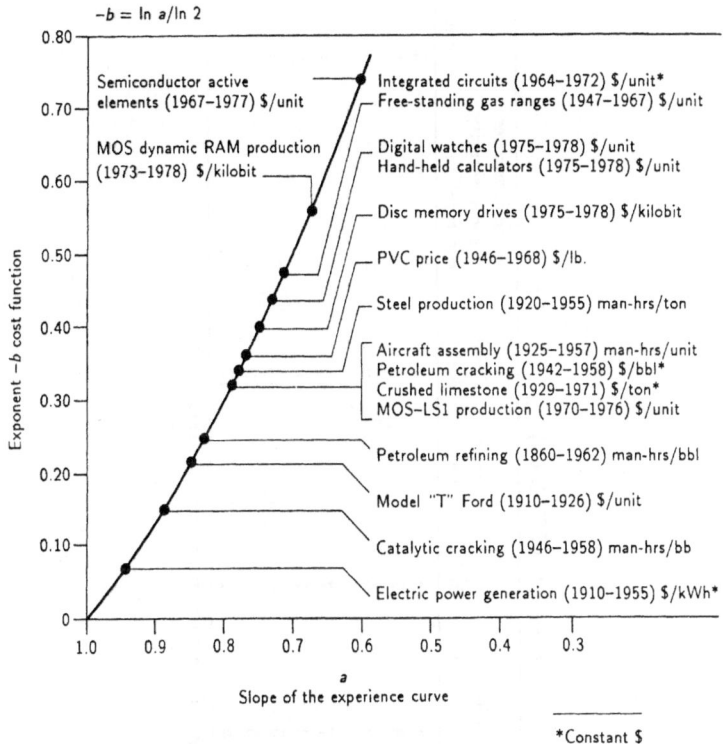

**Figure 3.5.** Experience curve parameters for various industries.

**Table 3.1.** Cost elements for small-, medium-, and large-scale manufacturing, in percent.

| Costs | Small-scale (stand-alone machine tools) | Medium-scale (FMS) | Large-scale (transfers lines) |
|---|---|---|---|
| Direct labor | 43.7 | 24.7 | 19.7 |
| Overhead | 13.5 | 13.9 | 23.7 |
| Capital | 17.8 | 33.1 | 29.8 |
| Other costs | 25.0 | 28.3 | 26.8 |
| Total | 100.0 | 100.0 | 100.0 |

Source: Wildemann, 1988.

Table 3.2. Industrial production cost structure, in percent: USSR (1985).

| Industry | Raw materials | Other materials | Fuel | Electric energy | Capital cost | Wage & salaries + social security contributions | Other |
|---|---|---|---|---|---|---|---|
| All industries | 63 | 4 | 4 | 3 | 9 | 14 | 3 |
| Food | 81 | 4 | 2 | 1 | 4 | 7 | 1 |
| Construction materials | 44 | 6 | 8 | 5 | 13 | 21 | 4 |
| Iron, steel, and non-ferrous metals | 58 | 7 | 7 | 6 | 11 | 10 | 1 |
| Machinery | 59 | 4 | 1 | 2 | 8 | 22 | 5 |
| Wood, paper, and pulp | 49 | 5 | 4 | 3 | 11 | 24 | 6 |
| Electro-energy | 3 | 6 | 54 | 1 | 23 | 10 | 3 |
| Fuel production | 53 | 5 | 1 | 5 | 17 | 13 | 7 |

Source: Narodnoye Khoziaistvo SSSR v 1985 g. – Financy i Statistica, Moscow, 1986, p. 126.

## 3.3   The Productivity Dilemma

It is significant that the history of manufacturing technology since 1900 is
essentially the history of progress down the cost curve in *Figure 3.4*. Each
successive stage in automation identified on the curve – with the exception of
numerical controls – occurred in historical sequence. When Ford Motor Co.
was first organized in 1903 most of the parts for Ford cars were supplied by
Dodge Bros. machine shop, which was perfectly prepared to switch on a moment's notice from gasoline engines to steam engines, or for that matter to
agricultural machinery (Galbraith, 1978). As a matter of fact, early models
of the Ford cars suffered from a variety of defects, including some rather major ones: "cooling system did not cool...brakes did not brake...carburetor
did not feed fuel to the engine" (ibid). There was even a case where the steering gear was installed backward. Yet, when these problems were brought to
the attention of Ford, they were quickly remedied and "the reputation of the
company suffered no lasting damage" (ibid). In effect, because the production technology was so flexible, the cost of a model change at that time was
virtually nil.

It is well known that Ford led the way to product standardization in the
automobile industry with the Model T, introduced in 1908. During the following two decades, Ford introduced many innovations in the manufacturing
process while keeping the product essentially unchanged. Of these, the best
known is the moving assembly line (c. 1916). New and faster machine tools
had also been developed – especially the "production grinder," which sharply
cut the machining time needed for complex engine parts such as crankshafts
and camshafts. Within the plant, however, the machines were still independent stand-alone units. However, the changes in Ford itself were mostly
organizational. Higher levels of production were made possible mainly by
standardizing the product and job specialization [following the prescriptions
of Frederick Taylor, the pioneer of "scientific management" (Taylor, 1911)].
Ford's basic objective, which was finally achieved around 1920 was to minimize machine setup time and eliminate hand "fitting."

By 1920, of course, Ford was producing cars on a very large scale, and
unit costs had dropped sharply. Still, when Ford converted from the Model
T to the Model A in 1926, the cost of that conversion was $25 million,
or roughly $17 for each car Ford sold in 1924. William Abernathy of the
Harvard Business School has estimated that the value of the model change,
to Ford, was $1.5 billion (1958 $), or about $500 million in 1926 prices. This
was some 20 times the cost of the change. By contrast, when Ford introduced

the Mustang in 1963, the conversion cost $59 million (1958 $), while the cost of introducing the Fairmont and Zephyr models in 1977 was $600 million (Galbraith, 1978). Meanwhile, the average value of the model changes to the firm in the postwar period, as estimated by Abernathy, declined to about $250 million (1958 $).

This made the Mustang well worthwhile as a new product, but all the model changes since then have been losers, at least based on Abernathy's evaluation methodology, as illustrated in *Figure 3.6*. The sharp rise in costs of model changes between the 1920s and the 1970s was almost entirely due to the introduction of "Detroit automation," as described above, during the intervening years. While unit costs are very low if the scale of output is large enough, the mass-production technology based on synchronous transfer lines is extremely inflexible. Virtually any change in the design of the car means a new plant has to be built. On the other hand, every such plant is unique. An auto engine plant built around 1980 with a capacity to produce 120 engines per hour, or up to 500,000 engines per year, would have cost at least $250 million, of which over half was for the machines themselves. The high cost of the machines is (or was) largely due to the cost of engineering and design. In an article for the 75th anniversary of the US Society of Automotive Engineers, Ralph Cross, Sr., chairman of the board of Cross & Trecker Corp., one of the largest machine tool manufacturers, explains the problem:

> Every spindle in every [drill] head must be rotated at the proper speed, and must be correctly positioned in its housing relative to the other spindles, and special gear trains must be provided to drive the spindles, etc. In addition to the high cost, the lead time to design and build the average multiple spindle head is usually a matter of several months (Cross, 1980, p. 39).

Cross continues:

> Roughly speaking, it takes 60,000 man hours of engineering to make the drawings for a dedicated manufacturing system that will produce cylinder blocks at the rate of 120 per hour. Practically all of this engineering work is done without the benefit of any mechanization (ibid, p. 40).

This is true even when making allowances for standardized drawings, CAD systems and computerized systems for making parts lists, and so on. This is a fixed cost that can be allocated only to one unit of output. Each plant is essentially custom-designed to produce a single family of car bodies (or a single engine model). Any change in the design of the car means a new plant has to be built. This is very expensive, precisely because the custom-

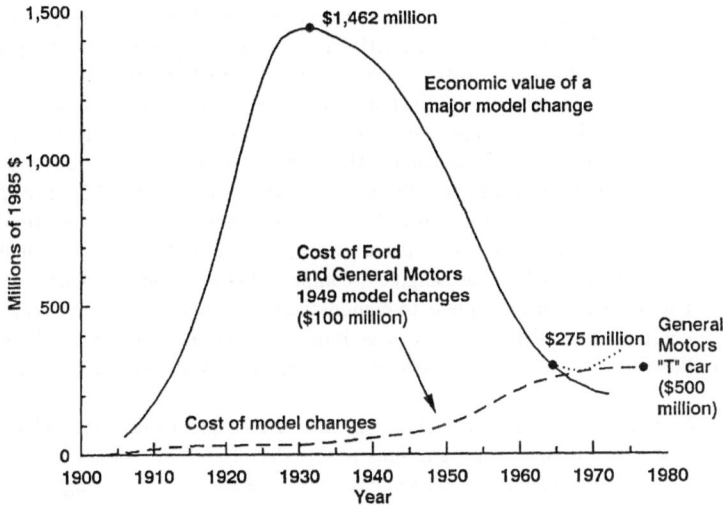

**Figure 3.6.** Value and cost of a major model change. Amounts in parentheses indicate current $. Source: Abernathy, 1978.

designed, one-of-a-kind dedicated plant is itself the product of labor-intensive job shops (see *Figure 3.4*).

Thus, economies of scale work both for and against mass production. Because the dedicated manufacturing plant is unique, its unit cost is the maximum possible for that product at that level of technology. The high cost of model changes in the automobile industry is directly related to this fact. If the cost curve for automobile engine plants is the same as the curve for engines, and if one could imagine producing a million copies of such an engine plant, its unit cost would drop to something closer to $1.5 million, give or take 25%. Obviously this is unrealistic and, to that extent, irrelevant.

However, it is not so unrealistic to imagine the following: suppose that the engine plant were put together from 100 flexible, multipurpose, programmable modules. (The term "flexibility" is used here in contrast to "specialization." The flexible module might be "hard wired" and used only for a single purpose for the life of the plant.) Suppose, too, that each standard flexible machinery module is itself produced in lots of 1,000 (using a

sophisticated flexible manufacturing technology somewhere in the middle of
*Figure 3.4*). In this case, based on the long-run cost curve (*Figure 3.4*) a
plausible estimate of unit costs for a batch of 1,000 would be 10% of the cost
if only one single unit was fabricated in a job shop. The fixed capital cost
of the engine plant assembled from modular units drops to roughly 10% of
its previous level. As a consequence, the unit cost of the engine drops too.
But in production runs of one million, the difference in unit costs is only of
the order of $250 per unit, or 10%. On the other hand, for production runs
of 10,000 engines, the fixed cost per unit drops by 90%, from (say) $25,000
per engine to only $2,500 per engine – comparable to the variable costs.

The point of this example is to demonstrate that modularization of capital equipment could have a very large impact on the unit cost of anything
produced in batches of (say) 10,000 or less. This point will be taken up again
later.

As pointed out earlier, engineering and design are fixed costs, which
must be paid whether or not anything is actually produced. In the case of a
dedicated auto engine plant, all the costs must be allocated to a single unit
of output – the plant. Given the massive capital investment for a new plant,
it is understandable that the firm wants to keep the plant operating for as
long as possible without change.

Bearing these points in mind, Abernathy has argued persuasively that
the inflexibility of the capital equipment accounts for the high cost of innovation, and hence the reluctance to innovate, that characterized the US auto
industry in the 1960s and 1970s. The slowdown in innovation, in turn, made
the industry less competitive and opened a "window" for Japanese competition in the late 1970s. The problem of slow technological change caused
by inflexibility was aptly termed the "productivity dilemma" by Abernathy
(1978). The conflict between current production cost minimization and the
cost of innovation is summarized in *Figure 3.7*.

## 3.4   Flexibility and Economies of Scope

The major thrust of John Kenneth Galbraith's (1978) book *The New Industrial State*, quoted earlier, was that large firms have so much market
power that market *forecasting* in a competitive environment is effectively
being replaced by *planning* in an environment controlled by the oligopoly.
Galbraith's observations shrewdly summed up the changes that had begun
in the US automobile industry after 1910. There is no doubt at all that

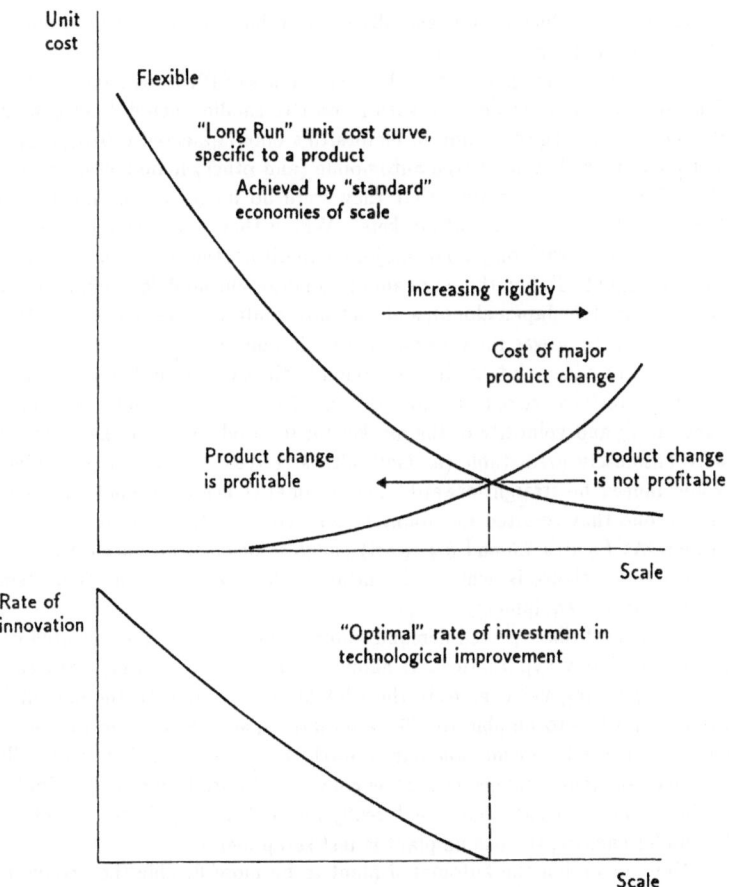

**Figure 3.7.** Productivity dilemma. Source: Abernathy, 1978.

the three dominant US automobile manufacturers constituted an effective oligopoly led by General Motors after World War II. Competition among the three firms was fierce but limited largely to marketing and styling. As noted above, major model changes were becoming rare because of high costs. Thus, for major components – like engines, which were shared among many models of cars – demand was actually quite stable and production could be planned for many years ahead.

All this began to change after the energy crisis of 1973–1974 and the sudden shifts in consumer demand with respect to smaller fuel-efficient cars.[2] By the time the third edition of Galbraith's book appeared (1978), events had made it irrelevant. Large automobile (and other) manufacturers were already beginning to realize that they were no longer competing in *national* markets but in *world* markets. Where they had formerly enjoyed oligopoly status with only a few major competitors, the situation shifted almost overnight. Today, there are about 25 major automobile manufacturers in the world (9 in Japan alone), and each now confronts a far more uncertain environment than was the case one or two decades ago.

From the viewpoint of simple economic theory, the problem is that a manufacturer's choice of technology depends (in a not so obvious way) on the uncertainty and volatility of the market for its product. If market demand were absolutely predictable (as Galbraith asserted), the planning problem would indeed be straightforward. The optimal choice of technology would be the one that resulted in minimum unit costs at the expected scale of output. As *Figures 3.3* and *3.4* clearly suggest, for mass-produced products the least-cost choice is achieved by means of hard automation: dedicated, special-purpose equipment.

But what if the expected demand suddenly evaporates? This eventuality is not completely hypothetical. It happened in the case of large heavy cars with gas guzzling V-8 engines in the USA after 1974. Ideally, the manufacturer would like to be able to offer alternatives, such as a 6- or 4-cylinder engine, a 5-speed transmission, and so forth. However, an engine plant built to make 8-cylinder engines *cannot* be converted to produce 6- or 4-cylinder engines. (This is what "dedicated" really means.) If there is no market for 8-cylinder engines, the engine plant is just scrap metal.

Why not design the automated plant to be more flexible (i.e., convertible)? That is possible, of course, but there are extra costs to design it that way. Many machines must have two different operating modes, for instance, and the gear-trains linking the work stations together in a transfer line must have extra gears. General Motors estimated informally that an automated plant capable of producing four models of engine would cost about 25% more to design and build than a plant dedicated to only a single model of engine.

In principle, firms always have the opportunity of adding *some* flexibility to their manufacturing system, at some premium cost. Or they can add *more* flexibility by decreasing fixed costs and increasing variable costs. The

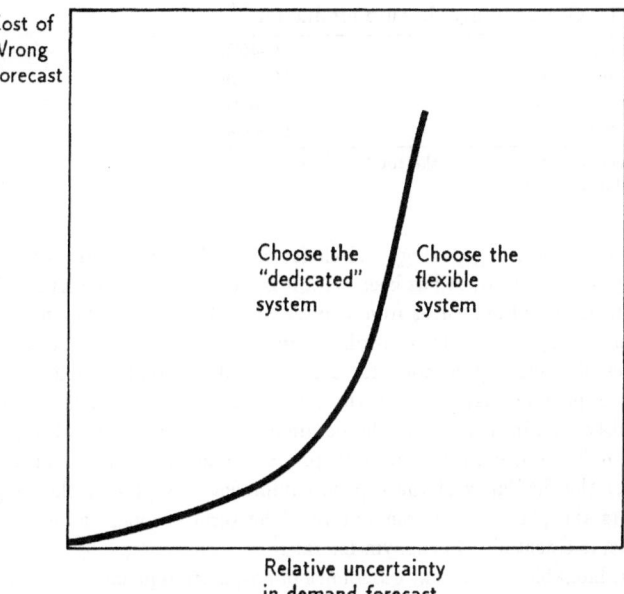

**Figure 3.8.** Value of flexibility in manufacturing. Source: Goldstein, 1981.

more uncertain the market, it seems, the more flexibility is worth (*Figure 3.8*). It is possible to reduce these relationships to quantitative terms using a simple cost model assuming that the fixed and variable costs for alternative systems are given (e.g., Goldstein, 1981). This is not the place for a detailed analysis, however.

It is sufficient to note one salient point. In the extreme case of a totally dedicated and totally automated engine plant, for instance, none of the capital equipment is convertible to other purposes. By contrast, in a conventional job shop with skilled machinists and general-purpose machines *all* of the capital equipment is convertible. Obviously there are intermediate cases. In fact, it is reasonable to assume a continuum, with capital convertibility (or adaptability) ranging from zero at one extreme to 100% at the other (*Table 3.3*). Capital convertibility is evidently one measure of manufacturing flexibility. It is now possible to analyze (and even, in principle, to

**Table 3.3.** Convertibility[a] in auto production.

| | |
|---|---|
| Transfer lines | 5–40% |
| Flexible manufacturing systems | 50–80% |
| Flexible assembly systems | 40–50% |
| Industrial robots | 60–80% |

[a]Value of equipment as part of the total instalment costs.
Source: Wildemann, 1988.

optimize) manufacturing strategy and choice of technology in terms of market uncertainty. From what has been said, it is clear the more uncertain the market, the greater the returns from capital convertibility (i.e., flexibility).

The increasing returns from flexibility in this sense have been denoted "economies of scope" by economists. The notion has broader applicability than may appear at first, because the intermediate case – partial capital convertibility – is, in fact, by far the dominant one. It is of critical importance in small- to medium-batch, multi-product manufacturing. Evidently, the greater the flexibility of the capital equipment, the greater the range of products the plant can accommodate. The term "scope," as used by economists, is essentially a synonym for range.

It is fashionable to say that economies of scope are replacing economies of scale in manufacturing. Insofar as this is just another way of saying that capital convertibility or flexibility is increasing in importance, one cannot argue with it. However, it is not by any means clear that there is a true substitution of a new growth mechanism for the old one. Economies of scale exist in both flexible and dedicated systems, although the returns to scale may be *quantitatively* somewhat greater in the latter case. Similarly, economies of scope can exist at any scale of production. The real issue to be resolved is the extent to which greater flexibility in manufacturing technology will have to be achieved *at the expense of* returns to scale. This topic is considered in Chapter 7.

An interesting sidelight on the productivity dilemma discussed in Section 3.3 is that flexibility of response to changes in the marketplace is much more important in manufacturing industry today than it was two decades ago. The introduction of new models of cars or refrigerators can no longer be viewed as an aggressive strategy that might upset the comfortable oligopolistic apple cart. It is, more realistically, a *defensive* strategy – a necessary investment in survival analogous to the replacement of obsolete capital equipment. But such a strategy is only feasible for firms with flexible technology and a capability for capital-sharing among a variety of products. In other words, rapid technological change is incompatible with maximization of returns to scale, but is perfectly compatible with maximization of returns to scope.

## Notes

[1] As noted in Section 2.2, practical methods of quantification of this variable are not yet widely known, although their development seems inevitable as CAD technology becomes universal.

[2] It is sometimes forgotten that each of the four US auto manufacturers hastily introduced new smaller models with a 4-cylinder engine in the early 1970s. Ford's Pinto was typical. All were briefly popular during and immediately after the Arab oil boycott, the resulting shortages and the sharp rise in prices of 1973–1974. Thereafter, they were conspicuous failures in the market, as the small imports from Europe and Japan were clearly superior in quality to the domestic product.

# Chapter 4

# Flexible Manufacturing Technology

## 4.1 Programmability

The idea of flexible manufacturing, used in this book, particularly in this chapter, refers to the use of programmable controls as opposed to mechanical linkages or "hard wiring." At first glance, perhaps, the shift may seem to be of minor importance even as compared with other new manufacturing technologies. For one thing, the degree of flexibility gained is relatively small, especially when programmable machines are contrasted with human workers, who are far more flexible than any machine will ever be in our lifetimes or our children's. Of course, the purpose of this book is to make the case that the change is not small and insignificant at all. On the contrary, it is revolutionary in its implications.

Where and how did the change begin? The critical prerequisite was undoubtedly the first electronic computer itself, built by a team led by J. Presper Eckert and John Mauchley at the Moore School of Engineering at the University of Pennsylvania. The project began in 1944 under military sponsorship, and the result, known as Electronic Numerical Integrator and Computer (ENIAC), was publicly unveiled in 1947. The developers of ENIAC went on to build the first commercial computer, UNIVAC I, which appeared in 1950. However, electronic computers in the 1950s were too expensive, too hard to program, and much too unreliable for most purposes. They were not directly applied to manufacturing problems until two more decades had passed.

The first step toward programmability in the factory was the development of so-called numerical controls (NC) for individual stand-alone machine tools, in the mid-1950s. Other complementary mechanical and electronic technologies have appeared on the scene and have changed since then. Semiconductors have made electronic devices smaller, cheaper, faster, and much more reliable. They have paved the way for microprocessors. Microprocessors and random-access memories on a "chip," in turn, permit "smart" sensors and adaptive machine-tool controllers that can react automatically to changing conditions.

The ability of machines to receive and interpret instructions, sense their environment, and communicate with each other (and with higher-level controllers) and to do it all electronically is of enormous significance for the future. Indeed, most other technological innovations are relatively insignificant by comparison. These developments have brought us up to the threshold of computer integrated manufacturing or CIM, where we stand today.

## 4.2   Numerical Controls before Computerization

Punched paper or magnetic tape controls for machine tools were conceived early in this century. The idea was pioneered as early as 1906 (by the Sellers Co.) and again in 1921 (E. Schleyer), but the time was not yet ripe. Nonprogrammable (cam-controlled) automats and mechanical indexing transfer machines have remained overwhelmingly dominant until recently, even though approximately 70% of all metal parts production is in small- and medium-sized lots that would seem suited to programmable machines (Cook, 1975).

The event that triggered the development of numerical controls after World War II was a Lockheed Aircraft Co. design featuring "integrally stiffened skins" – a new structural concept (1948). Up to that time the "skin" of an aircraft was made of sheet metal, bent and riveted onto a frame consisting of ribs and stringers assembled from individual pieces. The new concept proposed to use airfoil-contoured pieces with inner surfaces "pocketed out," leaving the equivalent of stiffeners and reducing the need for assembly and riveting.

John T. Parsons, owner of a machining company that specialized in such items as helicopter-rotor blades, conceived of a method of manufacturing the integral stiffened parts by feeding data by punched paper tape to a milling machine (*American Machinist*, 1977). The US Air Force sent a team to the

Parsons plant (Traverse City, Michigan) in late 1948 to evaluate the idea. This was followed by a machining demonstration at Snyder Corp., in Detroit, on a 16″ wing model tapering from a 6″ chord at one end to a 4″ chord at the other. All machine settings were computed mathematically. Then a random cross section was selected, and a matching template was produced, using the same calculations, on another machine. The two matched. This led to a research contract beginning in June 1949 for intensive development work that was carried out for the Air Force by the Servomechanisms Lab. at MIT. By 1951 an experimental system had been built, and application studies started. A report prepared by the MIT group in 1953 indicated that the system would be practical.

An independent effort at Arma Corp. by Fred Cunningham actually preceded the MIT work. Arma demonstrated a tape-controlled lathe in 1950, but dropped the effort when the Korean War broke out. Later, Arma encountered a problem producing non-circular gears for the Army T-41 Rangefinder. Cunningham modified a conventional gear shaper to accept a continuous stream of data (manually pre-computed), and used 16mm film and a photo-electric "reader" to control the machine. This machine was in commercial operation in 1952 producing precision gears.

Machine tools with NC controls were demonstrated and offered commercially in 1955. A sequence of tool positions and feed rates was specified via a punched paper on magnetic tape. The early controllers were expensive and (by modern standards) difficult to program.

An early partner (and outgrowth) of NC technology was the development of the so-called machining center (MC), first introduced by Kearney & Trecker in 1958. Machining centers are multi-axis NC milling machines with the addition of automatic tool-changing capability. Machining centers are therefore capable of carrying out a sequence of cutting operations on a single part, using up to 90 different tools. They are well suited for small-batch production of very complex metal shapes, e.g., for the aerospace industry. Machining centers are also central to the so-called flexible manufacturing cells (FMCs) and flexible manufacturing systems (FMSs) that have been widely adopted for batch production in the 1980s.

## 4.3 Computers and Microelectronics

Computers, microprocessors, programmable controllers (PCs), and "smart sensors," in the sense used above, are now entirely based on the technology of microelectronics. These are all essential prerequisites to advanced forms of flexible automation.

The first great breakthrough that made these modern developments possible was, of course, the development of semiconductor-switching elements (transistors) by John Bardeen, Walter Brattain, and William Shockley of Bell Telephone Laboratories in 1948. Semiconductors had the virtues of being shockproof, reliable, and energy efficient. In time, it also became clear that they could be made very small. The trend toward miniaturization (and, later, micro-miniaturization) has proceeded very rapidly, because of a "virtuous circle" of linked relationships.

Each reduction in the physical size of a circuit element results in a corresponding reduction in the electric power required, per unit operation. This, in turn, reduces the requirements for heat dissipation. This permits higher operating speeds and more compact circuitry. The performance of a computer, telephone switchboard, TV set, or radar navigation system tends to be closely related to the number of distinct circuit elements it embodies. On the other hand, the more elements there are the more interconnections there must be. It was realized already in the 1950s that manual assembly operations – and the resulting opportunities for errors and defective connections – would soon be the limiting factor in electronics.

Luckily, the "number's barrier" was broken almost as soon as it was recognized, by the second big breakthrough in 1959–1960. That was the so-called *integrated circuit* (IC), which combined transistors with other components (capacitors, inductors, resistors, etc.) composed of a multilayered stack of thin films deposited on an insulating ceramic substrate. This major invention is jointly attributed to Jack Kilby at Texas Instruments Corporation and Robert Noyce at Fairchild Semiconductor. (Noyce later left Fairchild to help found Intel Corp.)

The ICs of the early 1960s have been followed by several generations characterized by ever smaller individual circuit elements packed more and more tightly on a chip. The first generation (1960–1965) is sometimes called *small-scale integration* (SSI), referring to devices with up to 10 gates or bits of memory per device. The second generation (1965–1970) was medium-scale integration (MSI), characterized by 10–100 gates or bits of memory per device.

The third generation known as large-scale integration (LSI), arrived about 1970 with Intel's introduction of the first (1K) random-access memory (RAM) on a single chip in 1970. This was followed by Intel's 4-bit microprocessor in 1971. Very large-scale integration (VLSI) corresponds roughly with the

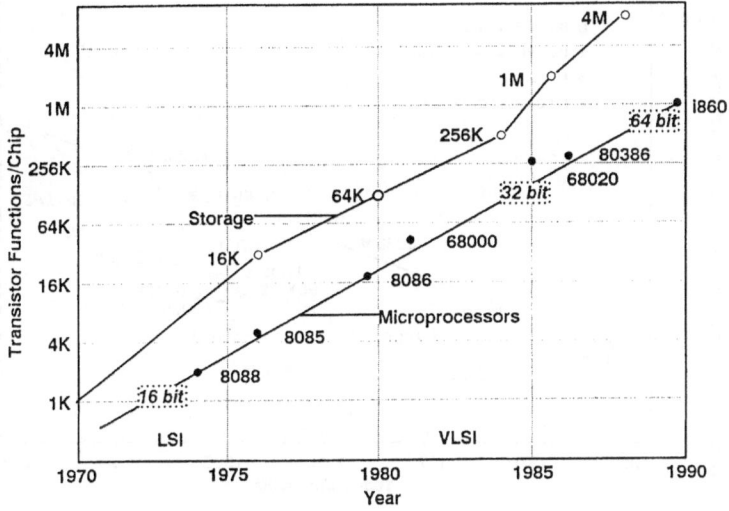

**Figure 4.1.** Development of memory and microprocessor chips. Adapted from: Bursky, 1983.

microcomputer on a chip and the 16K RAM (c. 1977), while ultra large-scale integration (ULSI) corresponds roughly with Western Electric's first million bit RAM (c. 1985). Progress has accelerated: in early 1987 NTT (Nippon Telephone & Telegraph) announced a generation-skipping 16-million-bit RAM chip; and in 1989 came Intel's 64-bit 1860 microprocessor, with 1 million circuit elements and the computational power of a supercomputer. The trend in circuit density (circuit elements per chip) since 1970 is shown in *Figure 4.1*. The computational efficiency of computers is, of course, directly related to the complexity of the circuitry. Trends in computer capability are shown in *Figure 4.2*. Precomputer technology is shown in *Table 4.1(a)*, and the first five generations of computer technology are summarized in *Table 4.1(b)*.

Unit costs (i.e., costs per gate or bit of memory) have moved down essentially in step with the number of elements per chip. Chips are manufactured nowadays by a complex but highly automated and capital-intensive process in which direct (i.e., hands-on), human labor plays almost no role. In fact,

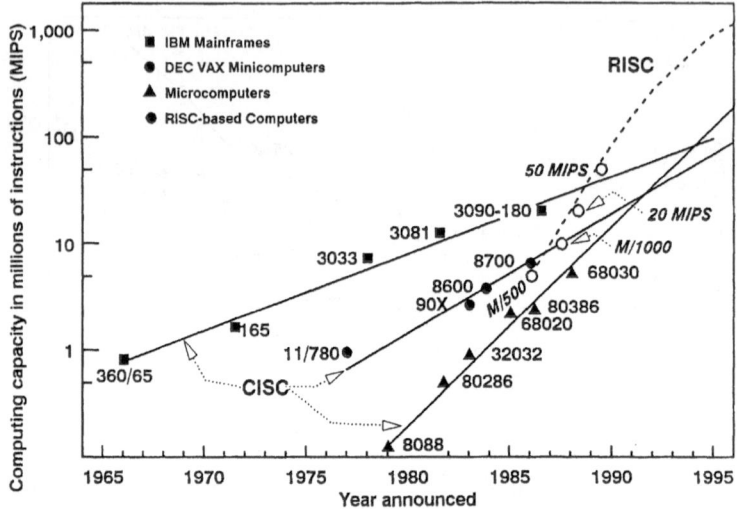

**Figure 4.2.** Efficiency of different computing architectures. Adapted from: *Electronic Design*, 1988.

modern plants humans must be rigorously kept away from the actual man-ufacturing steps because of the danger of contamination. By far the major elements of chip cost are now the design and the specialized capital equip-ment. The marginal cost of production is virtually the cost of materials only, which is negligible.

Price trends for logical functions are shown in *Figure 4.3* and for com-putation tasks in *Figure 4.4*. In relative terms, prices (and costs to users) have declined by a factor of about one million since the era of vacuum tubes. Impacts on system costs are summarized in *Table 4.2* and *Figure 4.3*.

Further technological improvements and corresponding cost reductions seem virtually assured by the enormous R&D resources currently being in-vested in microelectronics technology. A number of major extensions, includ-ing opto-electronics and organic chemical molecular (*molecutronic*) devices, now appear to be potentially feasible and perhaps fairly imminent.

**Table 4.1(a).** Pre-computer technology.

| | Pre-computer | |
| | to 1930 | 1930–1945 |
| --- | --- | --- |
| *Configuration of system* | Operator, mechanical calculator | Electro-mechanical relay calculator hard-wired program |
| *Logic hardware* | Gear wheels | Relays $4 \times 10(-8)$ |
| *High-speed memory hard. (short-term)* | None | |
| *Memory (mass)* | None | |
| *Data input mode* | Keyboard | |
| *Data output mode* | Mechanical display | |
| *Program, software systems* | NA | NA |
| *Typical tasks* | Log tables Astronomical tables Ballistic tables Navigational tables Scientific problems Gear shapes | |
| *Major entries* | Pascal, Leibnitz, Babbage, Monroe, Burroughs, Royal IBM | Bell Labs (Stibitz) Harvard (Aiken) Zuse 23 IBM-SSEC |

## 4.4 Computer Numerical Control of Machine Tools

The early 1970s saw rapid improvement in the basic technology of machine control due primarily to the introduction of microprocessors in 1969 by Intel Corporation. Microprocessors and pressure/torque sensor were successfully adapted to machine tools (and robots) in 1973–1974. The increased availability of computer power in the early 1970s permitted the introduction of far more flexible machine controls, known as *computer numerical control* or CNC. The advent of CNC also permitted another development: the

**Table 4.1(b).** Five generations of computer technology.

|  | First Generation | | Second Generation |
|---|---|---|---|
|  | 1946–1950 | 1951–1955 | 1956–1964 |
| *Configuration of system* | Electronic computer, stored program (CPU) Assembly language | | Symbolic programs |
| *Logic hard.* | Vacuum tube (CRT) $7 \times 10(-6)$ MIPS | $10(-4)$ MIPS | Transistors Hybrid ICs 0.2-0.4 MIPS |
| *High-speed memory hardware* | Mercury delay tube., Cathode ray tube (CRT) | | Ferrite cores |
| *Memory (mass)* | Punched cards | Magnetic tape | Magnetic drum |
| *Data input mode* | Punched cards, tape | Magnetic tape | |
| *Data output mode* | | Mag. tape, Print. cards | High-speed printer |
| *Program, software systems* | Programs in machine language | Assem. lang. IBM, Speedco' Univac Flowmatic' | Compilers FORTRAN 57 COBOL (59) ALGOL (60) |
| *Typical tasks* | Census tables, Actuarial tables Monte Carlo simulation Linear programming Matrix inversion | | Missile guidance, i.e., fire control Air defense General accounting functions, payroll Integer programming Econometric models |
| *Major entries* | ENIAC (Eckert-Mauchley) | Univac I IBM-701, 704, 709 IAD Whirlwind | IBM-7090 IBM-1401 Philco 212 RCA 300 |

**Table 4.1(b).** Continued.

| Third Generation | Fourth Generation | Fifth Generation |
|---|---|---|
| 1965–1974 | 1975–1982 | 1982–1990 |
| Time-sharing<br>terminals<br>Networks | Hierarchies<br>Parallel processors<br>Net. "smart" terminals | Hierarchies of<br>computers linked<br>by LANs |
| Integrated circuits (ICs)<br>LSI<br>(MOS) 100 MIPS | 500 MIPS<br>VLSI<br>Circuits | VLSI systems<br>Graphics |
| LSI, RAMs, PROMs | Semiconductor dynamic<br>random-access<br>memory (DRAM) | Read-only Memory<br>ROM compact disks |
| Hard disks (large) | Magnetic disk (floppy, hard)<br>Hard disks (small, mid)<br>Winchester drive | |
| Modems<br>VDU, tape | Smart terminals<br>VDU, "Mouse" | Graphics<br>Work stations |
| Ink-jet printers,<br>dot matrix | Display on CRT<br>Laser and LED printers | GDU<br>Plotters, printers |
| Basic time-<br>sharing system<br>Prolog, Pascal, LISP | C Language, UNIX<br>Computer Conferencing<br>Software packages for PC | Modular programs<br>Object-oriented<br>language |
| Graphics, Time-sharing<br>Large-scale simu.<br>Weather forecast.<br>Distributed processing<br>MRP (I) systems<br>Industrial process control<br>First gen. (CAD/CAM) | CAD, Networking<br>Text processing<br>Computer-enhanced recog.<br>Database management<br>MRPII systems<br>Commercial CAD<br>Scheduling systems | Vision processing<br>Expert systems<br>CIM systems<br>Graphic simulation<br>Advanced CAD<br>Rule-based factory<br>management, AI |
| IBM-360, 370<br>Burroughs 6500, 7500<br>CDC-3600, 6600, 7600<br>UNIVAC-1103, 1107, 1109<br>ILLIAC IV, CDC-STAR-100<br>STARAN-IV, DEC-PDP | Microprocessors<br>(intel) CRAY-1<br>Minicomputers<br>DEC PDP 11, IBM 30xx<br>Super-minis, DEC-VAX<br>Tandem, PCs (Apple) | Sun<br>Cray III<br>Micro-VAX |

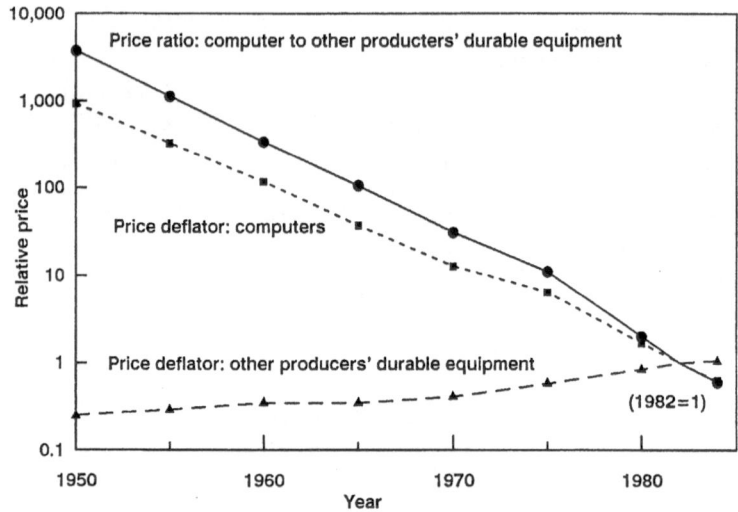

**Figure 4.3.** Declining relative price of computers. Derived from: Gordon, 1987.

simultaneous control of a number of independent stand-alone NC machines by a single computer (known as direct numerical control, or DNC). So-called *adaptive control* units for machine tools, based on pressure/force sensors in the work head to detect early signs of tool wear or misalignment, were first marketed as early as 1972–1973. This technology, still primitive, does not require CNC but obviously benefits from it.

Moreover, parallel developments in software development were permitting instructions to be given in more functional terms. Ideally, it should be possible for a machine or a machining cell to be instructed in natural language: e.g., "make 10 copies of model #XYZ 123." The supervisory computer would then consult an on-line data base to determine what parts are required and how many of each. It then consults a file of existing programs, decides – based on a *scheduling algorithm* – which machine or cell is to make the part, calls for the needed stock or other requirements from inventory, down-loads the instruction program for each desired part number to the microprocessor controlling the designated machine (or machining cell), and activates it.

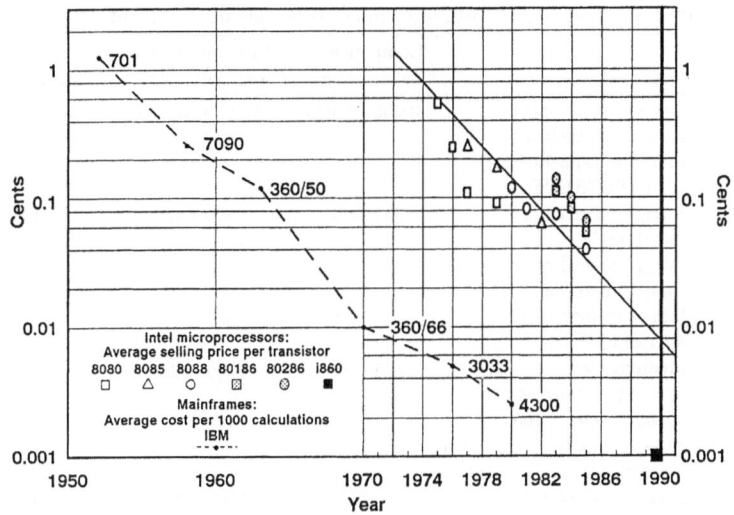

**Figure 4.4.** Cost of computation.

Modular program packages became available in the 1970s which cut programming time for CNC systems by a factor of three from 1971 to 1974 alone. The trend toward user-friendly software has continued. So-called fourth generation languages of the 1980s – exemplified by FOCUS, MARK V, RAMIS, IDEAL – are far more user-friendly than COBOL or FORTRAN, the assembly languages of the 1960s.

Perhaps partly as a result, the average cost of CNC machine tools purchased actually stopped declining in the early 1970s (*Figure 4.5*). This corresponds to the increased use of CNC in larger-scale production applications (requiring bigger machines) and, especially, a growth in the use of machining centers.

CAD is a system used for design purposes; it consists of a computer with a high-resolution display screen, a plotter, and sophisticated software for creating graphics and for combining and recombining graphic elements. The system is also linked to a library of previous designs, and may include rule-based programs for checking the conformity of design changes to specifications. Computer aided design (CAD) had its beginnings in proprietary systems

**Table 4.2.**  Cost impacts of major microelectronic developments.

| Evolutionary step | Components to assemble | Component and assembly costs[a] | Cost ratio |
|---|---|---|---|
| 1. Discrete-component systems (transistors, resistors, capacitors, etc.) DISCRETE | 20,000–30,000 | $6,000–$9,000 | – |
| 2. Integrated circuits (small-scale integration – less than 10 gates or bits of memory per device) SSI | 350–500 | $600–$900 | 10:1 |
| 3. Medium-scale integration (adders, counters, etc. – 100 gates or bits of memory per device) MSI | 125–150 | $250–$450 | 20:1 |
| 4. Large-scale integration (micro-processors and custom LSI circuits – more than 100 gates or bits of memory per device) LSI | 7–10 | $100–$200 | 50:1 |
| 5. Single-chip micro-computer VLSI | 1 | $5–$10 | 1,000:1 |

[a]Excluding backplanes, cables, cabinetry, etc. Source: NIRA, 1985.

invented by large aerospace manufacturers such as McDonnell–Douglas and Boeing. These early systems used mainframe computers, and their descendants today still do.

However, in 1970 ComputerVision Corporation successfully introduced the first complete "turnkey" CAD systems to the market. Rapid technological progress in micro-electronics has continued to improve performance and bring user costs down. *Table 4.3* lists the progressive complexity of CAD applications with increasing emphasis on "expert systems."

Roughly speaking, CAM systems are high-level supervisory systems that may carry out planning and scheduling functions for a plant and generate programs for individual machine tools or cells or both. Presently, CAD and CAM are largely separate, but it is clear that as designs (and design

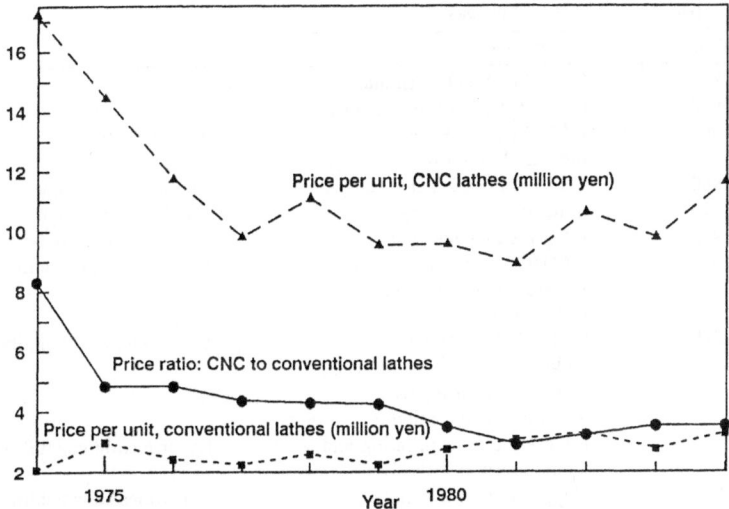

**Figure 4.5.** CNC and conventional lathes in Japan, units purchased and price ratio. Source: Jacobsson, 1986.

changes) are increasingly digitized the blueprint stage will eventually be by-passed. It is this linkage that is anticipated by the combination CAD/CAM. This acronym stands for *computer aided design/computer aided manufacturing*. These phrases are very nearly self-explanatory, except perhaps that it is unclear why "numerical control" (NC, CNC, or DNC) becomes CAM. Moreover, the detailed planning of a manufacturing process (e.g., a sequence of steps), starting from a set of design drawings and specifications, will increasingly be automated.

CAM has not developed as rapidly as CAD as an independent class of packaged and standardized software, since its applications are inherently more diverse. Moreover, there is no significant group of independent users (analogous to architects and design firms) that are not also manufacturers with proprietary products and processes to protect. Hence, most of the work in this field is undoubtedly in-house software development for specific applications. It is likely that the expansion of CAM applications is keeping pace with CAD. However, until CAD and CAM are truly linked into one

**Table 4.3.** CAD technology.

| Year | Technology | Capability |
|------|-----------|------------|
| 1961 | CAD, 2D, Single terminal | |
| 1963 | CAD, $2\frac{1}{2}$D, Multiple terminals | Mainly drafting |
| 1966 | CAD, 3D, Full-scale industrial applications | Mainly drafting |
| 1968 | CAD, Finite Element Analysis | Emphasis on design |
| 1970 | Simulation capability | Emphasis on design |
| 1972 | Integrated engineering | Experimental capability |
| 1974 | CAD/CAM, Bill of materials, Integrated engineering, and Manufacturing | Experimental capability |
| 1978 | CAD/CAM Networks, Online integrated engineering, and manufacturing with dynamic configuration | Experimental capability |
| 1982 | Integration of engineering and manufacturing with MIS | Experimental capability |
| 1984 | 3D geometric modeling | Experimental capability |
| 1986 | Integration with DSS | Experimental capability |

Source: Chorafas, 1987.

system, the dream of "industrial boutiques" producing "parts on demand" will not be realizable.

CAM systems provide computerized control of the actual manufacturing process for a part or product at the individual machine level. In more sophisticated cases, the system may control a group of machines linked together in a *cell* or *flexible manufacturing system* (FMS), to produce a family of geometrically related parts. In all but a few special cases, CAD and CAM systems are independent of each other, although the advantages of linkage are obvious.

In effect, the NC and CNC machine tool has become more flexible and able to take over more and more decisions, beginning with automatic actuation and "stop" conditions. An obvious extension of capability is toward automatic tool changing. Whereas in a dedicated machining cell or hard-automated synchronous transfer line tool changing can be scheduled in advance by the designer, this is not the case where a group of multipurpose machines is producing a wide variety of products. In the latter situation, it is essential that records of each tool's use (material being cut, speed, cutting time) be stored in memory. Moreover, the information must be available at

**Table 4.4.** Comparison between methods for machining an electric motor housing.

| Measure | Conventional machines | Machining center with automatic tool changer |
|---|---|---|
| Setups | 5 | 1 |
| Tools | 22 | 18 |
| Operations | 111 | 111 |
| Machines | 4 | 1 |
| Time (min) | 99 | 41 |

Source: Carter, 1982.

the scheduling level of a hierarchical control system – a level higher than the machine control level – so that individual machines are not shut down for tool changing in the middle of production runs.

One strategy to minimize such difficulties in multi-product shops is to incorporate into a single machine tool all the machining operations needed for a given part (e.g., milling, drilling, boring, facing, threading, and tapping). Extremely general-purpose machines, called *machining centers*, were introduced in the 1960s explicitly to exploit the capabilities of NC. Such machines may have as many as 90 different tools and programmable tool-changing capabilities. A typical comparison is shown in *Table 4.4*.

But the strategy of increasing the flexibility of individual machines is limited by the fact that only one of the 90 tools can be used at a time. Thus, the more tools and degrees of freedom the machine has because they might be needed, the more its maximum capability is likely to be underutilized during any particular operation. Machining centers are therefore mainly used for small-batch production of very complex prismatic parts like the motor housing cited in *Table 4.4*.

Evidently military needs played a major role in the innovation of numerical controls and CAD/CAM systems, primarily in connection with the production of very complex shapes such as airfoils. The US Air Force has continued to promote the technology with its technological modernization (TECHMOD), manufacturing technology (MANTECH), and integrated computer assisted manufacturing (ICAM) programs, which effectively subsidize the development and experimental and prototype installation of advanced manufacturing technology in the plants of aerospace contractors.

It is surely no accident that the aerospace companies in particular were among the first to employ the technology on a reasonably large scale, and remain among the leaders in its application. Indeed, NC tools were first used for very complex part shapes (rather than for small-batch production of simpler shapes). This fact may well have contributed to an impression that the technology was only suitable for very sophisticated users. If so, it probably deterred other potential users, especially among smaller firms. Thus, the original military sponsorship may have actually slowed down civilian adoption.

## 4.5   Robots

The introduction of robots, in contrast, owes little or nothing to military needs. The principal inventor of the underlying control technologies was George Devol, who worked at Remington-Rand in the 1950s. The Devol patents were purchased as a group by CONDEC Corp., which undertook to develop a commercial robot, the "Unimate." The principal entrepreneur and innovator for the next 20 years was Joseph Engelberger, who has been called the "father of robotics."

Industrial robots with point-to-point controls for simple material-handling tasks were first introduced commercially in 1959, and the first robot with path-control capability appeared in 1961 (the Unimate). Early industrial robots (1960s) were crude pneumatic or hydraulic heavy-duty manipulators with considerable flexibility in terms of degrees of freedom (up to 7), but they did not have particularly high levels of accuracy or repeatability. Robots with many degrees of freedom are inherently difficult to control because of the coordination problem. (Current robots tend to have fewer degrees of freedom for this reason.) However, early US robots were programmed and controlled by *analog* – rather than *digital* – means. This limited their usefulness essentially to stand-alone applications such as handling hot castings, spray painting or torch cutting (Engelberger, 1980).

However, the integration of robots into systems with machine tools and other equipment, or into assembly systems, was very difficult until the second generation of more accurate and reliable electric, digital (CNC) robots appeared in the late 1970s. Most robot manufacturers still make it hard to integrate their robots with other machines under higher-level computer control by retaining secret proprietary operating systems. However, robots of the 1980s are substantially more accurate and better coordinated (e.g., two robot hands can work together) than was the case for robots of the 1960s.

As noted above, the early US robot pioneers were slow to adopt the newer digital control technology, and many of those pioneers have since fallen by the wayside. The Japanese first saw the advantages of standardized, low-cost CNC machine tools and compatible CNC robots that are able to work with them under a common higher-level computerized control system. This almost certainly is the major reason for the large number of robots used today in Japanese factories. With greater accuracy, spot-welding and arc-welding applications became feasible, and indeed accounted for most industrial applications until 1983. The first practical assembly robots appeared only after 1980; they began to find significant applications during the present decade.

Robot capabilities are progressing steadily, but not spectacularly, primarily through improvements in controls and ease of programmability. A recent breakthrough in flexible robot gripper design promises to reduce the amount of specialized engineering needed for each application. Electric motor drives are replacing pneumatic and hydraulic systems for robots requiring greater precision, such as assembly. Operating speeds are increasing, but not dramatically. Robots, in general, work at about the same rate as humans. Their economic advantage is greater reliability and timelessness. In principle, robots can operate 24 hours a day – although this capability is seldom fully exploited.

However, so far as robots are concerned, the major technical breakthrough of the 1980s is the addition of vision and tactile sensors and feedback control to robots. Vision and taction are discussed in Section 4.8.

## 4.6  Flexible (Batch) Manufacturing: FMC and FMS

So-called *flexible manufacturing systems* or FMS have attracted much attention since the first attempt to combine several CNC machine tools with an automated materials-handling system under computer control (c. 1967) at the Molins Co. (UK). In an FMS, workpieces of different types (usually palletized) travel between and are processed at various programmable, multipurpose machine tools – mainly machining centers – and other work stations. An illustrative example is shown diagrammatically in *Figure 4.6*.

To integrate a number of machines or cells, a hierarchical top-down control system is often advocated (as illustrated in *Figure 4.7*). The reason is that detailed instructions for the actions to be performed by each particular machine must be converted into "machine language" specialized to

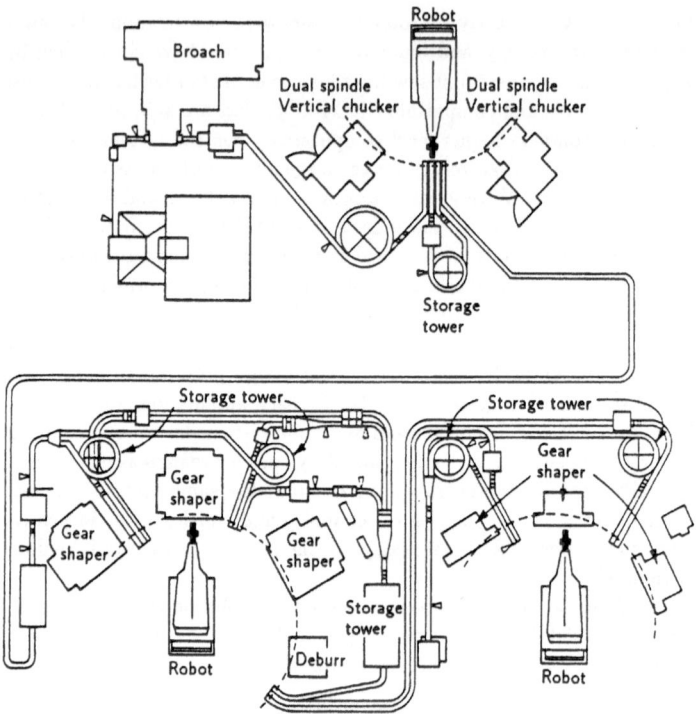

**Figure 4.6.** Flexible manufacturing system for pinion gears.

that machine interface. However, for efficiency, it is necessary for production engineers to be able to program all design, coordination, and scheduling functions in a single higher-level language. Thus, a supervisory computer should be able to translate from the high-level language used by the human engineers to the detailed machine-level languages understood by each machine. Increasingly, however, it seems likely that the control functions will be distributed around the system, rather than centralized in a single supervisory computer. Some implications of this will be considered later.

Parts flow through the system according to individual processing and production requirements. Both the materials-handling system and the machine tools are under automatic computer control, with minimal human

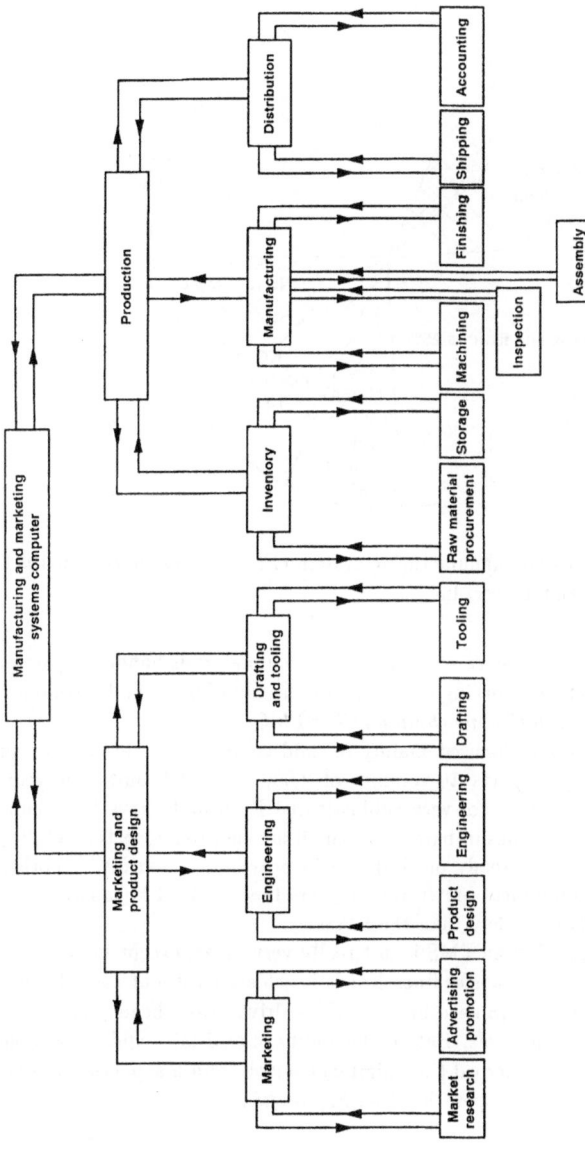

**Figure 4.7.** Computerized integration of manufacturing. In a completely computer aided design/computer aided manufacturing operation, there will be hierarchies of computers. Thus, the information and control loop from any one point in the operation to any other point will be easily facilitated. Source: *Modern Machine Shop 1984 NC/CAM Guidebook.*

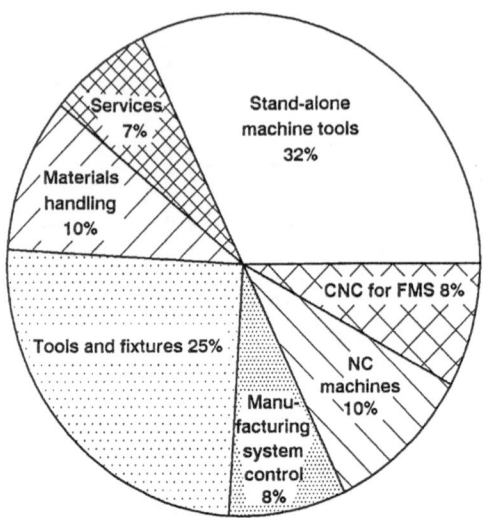

**Figure 4.8.** Flexible manufacturing system (FMS) hardware cost in 1984. Data: Kearney and Tecker, Inc.

supervision. In the absence of a programmable materials-handling system, a cluster of machines working on a common family of parts under common control is called a *flexible machining cell* or FMC.

Early applications focused mainly on mid-volume batch production of moderately complex parts at volumes of 2,000 to 50,000 units per year. More recently, a number of very sophisticated FMS units have been built to produce parts families in batches as small as one (assuming the detailed design drawings are already on file) and in near-infinite variety. Another recent trend is simplification: in recent years the simpler FMCs have been adopted much more widely than the FMSs.

The "flexibility" of an FMS is not really very great, except in comparison with a dedicated transfer line. Nor is it achieved without cost. Both a transfer line and an FMS need basic machine drives, work heads, materials-handling system, and tools. But the flexibility of an FMS requires variable speeds and cycles, numerical (i.e., digital) controls, and a supervisory computer to coordinate cell operation (see *Figure 4.8*).

**Table 4.5.** Cost of machine tool controls in FMS.

| Control capability | Cost ($ \times 10^3$) |
|---|---|
| Fixed sequence | $100 \pm 25$ |
| Variable sequence | $110 \pm 25$ |
| NC (Tape) | $125 \pm 25$ |
| CNC | $150 \pm 25$ |
| Adaptive, with sensing | $175 \pm 25$ |

In addition to the added hardware cost of an FMS is the cost of the system's software and the specialized programs need to implement a particular task. In a more sophisticated FMS with automated inspection or adaptive control capabilities, the cost of sensors and vision (or tactile) information processing must also be included. Expressing this cost breakdown as a relationship between cost and control capability, it is clear that the implemented cost increases as the level of control increases (*Table 4.5*). Numerical control (NC) capability adds about one-third to the per spindle cost of a typical machine tool, and the provisions for integrating CNC into an FMS add another 20%, roughly.

This cost comparison is only meaningful if we compare equipment manufactured on the same scale of outputs. Relative costs, too, will change over time. Many control-related components of flexible manufacturing systems are rapidly dropping in price, as pointed out earlier. As the price of these components decreases, so will the cost of the FMS. The net result of falling costs and increasing complexity of computers and NC machine tools is likely to lower the hardware cost of flexible manufacturing (FMS and FMC).

A point made rather emphatically in Chapter 3 was that the hardware cost of flexible factory automation might be cut sharply (perhaps as much as 10-fold) by effectively using *standardized* multipurpose equipment modules that could themselves be manufactured in much larger batches. These multipurpose modules will necessarily be equipped to operate at *variable speeds* and *cycles*, and they must be *electronically controllable*.

Here the essential difference between small-batch manufacturing in a multi-product plant and large-scale or mass production of a single product is reflected at the equipment level. In small-batch production (job shops), there is no need to synchronize the operations of different cells. Coordination can be crude, because no run is very long and workpieces in process can normally wait until a suitable machine becomes available for the next operation. Machine utilization can be increased at the expense of work-in-progress inventory, and vice versa. The optimum balance is determined by experience, or with the help of scheduling models. But machine utilization is likely to be quite low, and inventory of work in progress is likely to be high even in a well-managed job shop. Idle machines or exceptional delays should clue shop schedulers to modify normal processing sequences. When such problems are persistent, the remedy may be to add an additional stand-alone machine, or possibly to eliminate one that is unnecessary.

In a hard-automated, large-batch (mass) production environment, however, only one product is being made at a time, and the sequence of operations is fixed. Here, the ideal situation is one where the work-in-progress inventory is, essentially, one workpiece per work head. In principle, machine use is very nearly 100% when the plant is operating except for setup periods, breakdowns, and tool changes or other scheduled maintenance. Of course, a breakdown at any point in the fixed sequence causes the whole line to stop. In an imperfect world this limits the number of machine operations that can be linked safely in sequence without a buffer. Such a linked set of machines constitutes a *cell* in the mass-production environment.

The generic large-scale FMC or FMS will therefore consist of a number of "islands of automation" buffered by intermediate storage, but operating synchronously on average. The target operating mode would be such that the number of workpieces stored in each buffer unit fluctuates around half of its maximum storage capacity.

It can be assumed that each machine is controlled by a microprocessor which, in turn, communicates with a minicomputer at the cell level. The machine microprocessor contains a stored program of instructions for the machine, down loaded from the cell controller. Sensory automation monitors performance in real time. Any deviation from the expected status of the machine/workshop during processing would trigger a slowdown or even a "stop," which is signaled to the cell controller.

The cell controller coordinates materials-handling functions within the cell and provides the rhythmic "beat" that synchronizes the individual machine programs (such as a conductor synchronizes the musicians in an orchestra). Again, sensory feedback data monitor cell performance in real time, and deviations from the norm can result in a programmed shutdown of the cell and an automatic maintenance call.

The cell controller, in turn, communicates directly with neighboring cells in a *distributed control* scheme, or with a higher-level *supervisory computer* that coordinates other cells and buffers, as well as overall materials-handling functions. If one cell is down, the supervisory computer may instruct neighboring cells to continue to function temporarily, taking workpieces from buffer storage or feeding them into buffer storage.

In a very sophisticated FMS, there may also be several cells, in parallel, carrying out the same sequence of operations. In this case the supervising computer might bypass one cell and temporarily speed up the others to compensate. This would increase the rate of tool wear and result in earlier tool changes in the affect cells, but this would often be cheaper than simply reducing production for the plant as a whole.

Evidently, the computerized operating system for an LS/FMS in large-batch production mode would be quite complex, though qualitatively different from the operating system for a multi-product "parts-on-demand" plant. In many respects, the control problems are similar to those encountered in a traffic-flow network or continuous process plant, i.e., the buildup of nonlinear transients resulting from feedbacks in the system. The analogy between traffic flow and parts flow – in terms of phenomena such as collisions and congestion – is quite close.

The first generation FMS systems were largely custom designed to produce a family of parts in small- to medium-batch sizes. Once built, they were not particularly adaptable to other sizes or shapes. However, as adaptive machine control technology becomes increasingly practical in the 1990s and machine control software packages become more powerful and easier to use, more and more new, and virtually unmanned (second generation) plants will be built to make products that are less standardized and still subject to frequent design change. *Table 4.6* shows the average characteristics of several hundred FMS for which data are available at IIASA. A measure of technical complexity ($TC$) has been developed for purposes of analysis:

$$TC = 0.7MC + 0.35NCMT + 0.3IR + 0.6AGV + 0.3MH \ , \qquad (4.1)$$

where $MC$ refers to the number of machining centers, $NCMT$ is the number of other $NC$ machines, $IR$ is the number of industrial robots, $AGV$ is the number of automated guided vehicles, and $MH$ is the number of conventional materials-handling devices (such as conveyers). As shown in *Figure 4.9*, complexity increased from the early 1970s through 1986, but began to decline thereafter. *Figure 4.10* shows that installed cost can be fairly well explained by technical complexity for most countries. US costs are anomalously high, but this is largely explained by the artificially overvalued US dollar that prevailed during the 1980–1985 period when most of the FMSs were installed. Taking revised exchange rates, characteristic of more recent years, the US cost figures would be comparable with the other countries. It

**Table 4.6.**  Average FMS characteristics by area of application.

| Indicators | Metal cutting | Metal forming | Welding & assembly | Overall average |
|---|---|---|---|---|
| Number of NC machines | 6.9 | 3.9 | 18.0 | 7.1[a] |
| Number of robots | 3.1 | 1.7 | 29.0 | 6.8 |
| Technical complexity (TC) | 4.4 | 1.9 | 9.4 | 4.6 |
| Operation rate | 2.7 | 2.3 | 2.4 | 2.6 |
| Number of unmanned shifts | 1.0 | 0.8 | 1.6 | 1.0 |
| Number of product variants | 163.0 | 1138.0 | 88.0 | 216.0 |
| Batch size | 207.0 | 71.0 | 324.0 | 188.0 |
| Installed cost | 5.7 | 3.1 | 5.9 | 5.6[b] |

[a]6.9 for 309 FMS in 1984–1985 (ECE, 1986).
[b]7.1 for 98 FMS in 1984–1985 (ECE, 1986).
Source: IIASA.

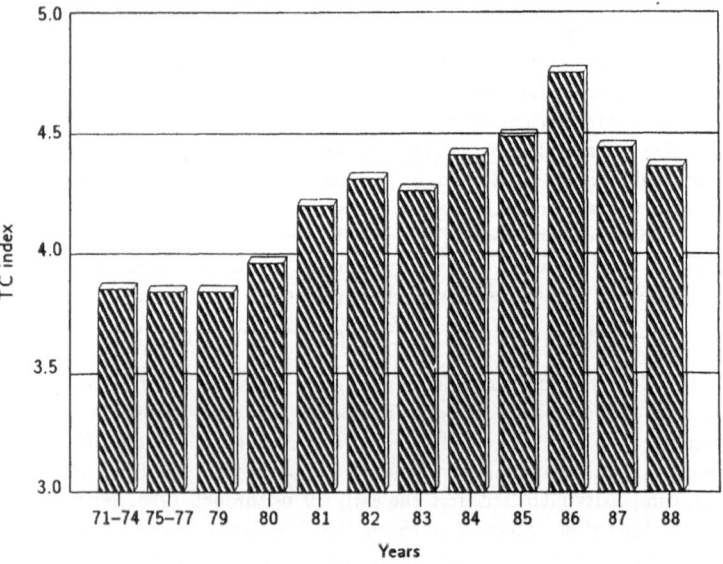

**Figure 4.9.**  Technical complexity of FMS. Source: IIASA.

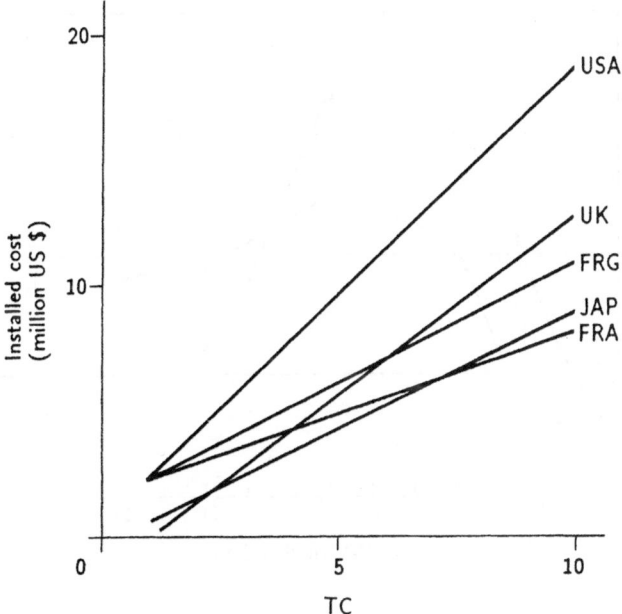

**Figure 4.10.** Installed cost vs. technical complexity.

is quite interesting to note that pay-back time is a parabolic (inverted U) function of both installed cost and technical complexity (*Figure 4.11*).

## 4.7   Adoption Strategies for FMS

As suggested above, there may be as many as three distinct classes of FMS, depending on the nature of the firm's business and its strategy. These "typical" cases can be defined briefly as follows:

(A)  The firm is a job shop, with very high flexibility and low volume (capacity). It is prepared to decrease its flexibility somewhat, in exchange for sharply increased capacity (5X). It tries to do this by increasing its fixed capital investment (2X), while holding labor ($L$) constant, but this is justified by increasing both labor and capital productivity (5X and

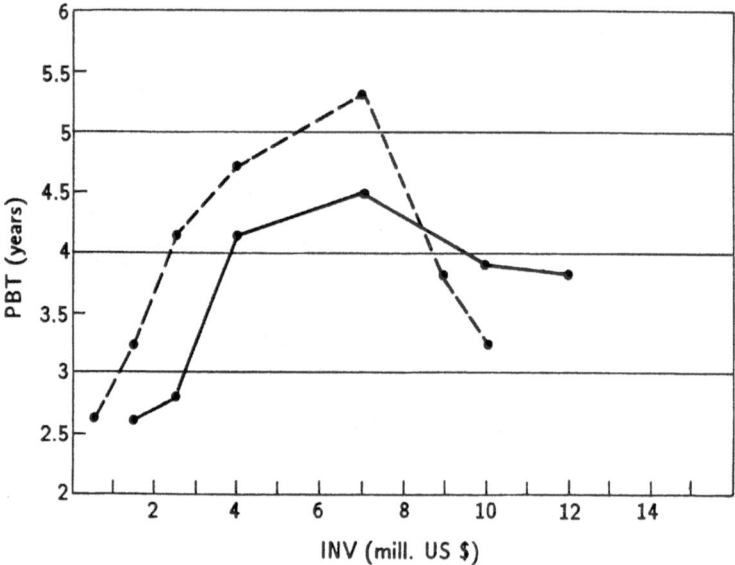

**Figure 4.11.** Pay-back time vs. installed cost. Source: IIASA.

5/2X, respectively). Other benefits include shorter lead times, reduced inventory, and better quality.

(B) The firm is a medium-scale producer with some automation and moderate flexibility. It attempts to increase both capacity and flexibility modestly. It tries to do this by holding labor constant, cutting the number of machines by two-thirds, and the fixed-capital investment by one-third. Apart from increased flexibility, it gains modestly in labor and capital productivity, plus lower inventory and shorter lead times.

(C) The firm is a conventional mass producer with very little flexibility. It tries to increase flexibility significantly, while holding capacity more or less constant. It also attempts to cut the number of machines by four-fifths, the number of (direct) workers by two-thirds, and the fixed-capital investment by one-third. If successful, it will increase labor and capital productivity by factors of 3.5X and 1.75X, respectively, while also enjoying other gains in lead time and inventory reduction.

These three cases illustrate different idealized situations. The potential benefits differ according to the situation, but so do the risks – the intangibles that are hard to reduce to firm financial figures. For case (A), for instance, the company must double its capital investment and try to sell five times as much output as before. This is no easy task for a typical job-shop producer with very limited marketing capability. On the other hand, there are lucky instances where increasing demand was the basic "driving force" for making the change. Of course, if the job shop is a service division of a larger firm, it may be able to increase its sales to other divisions of the firm without excessive difficulty.

Cases (B) and (C) are less risky, on the surface. Case (B) is primarily a reorganization of production to cut costs and improve quality, without major changes in either output or patterns of demand. It would be carried out as part of a long-range strategy leading toward CIM. Case (C) is probably a response to changing and less-predictable markets. The market is not growing fast, but product lifetimes are shorter and the company has to be able to respond fast. Also, quality control is increasingly important. These are the driving forces. Lower labor costs are welcome, but of secondary importance.

Management attitudes and unexpected software requirements are intangible risks in all cases, but perhaps most of all in case (C), where the management is probably conservative and resistant to change. On the other hand, case (C) can afford some miscalculation, whereas for case (A) there is probably no room for error. Fortunately, the management of case (A) is probably more flexible than the others, and hence less likely to make certain kinds of mistakes (see Chapter 6). *Figure 4.12* illustrates the three cases.

## 4.8   Sensors and Adaptive Control

Machine control requires sensory feedback. In the traditional machine shop and even in a "hard-automated" plant, the senses of experienced human machine operators and supervisors provide the sensory data (*Figure 4.13*). Gradually, machines and systems have been designed to use sensory data from other devices as indicated in *Table 4.7*. However until recently, most sensory devices were essentially one-dimensional, in that the information provided was a discrete sequence of some measure, such as pressure or voltage.

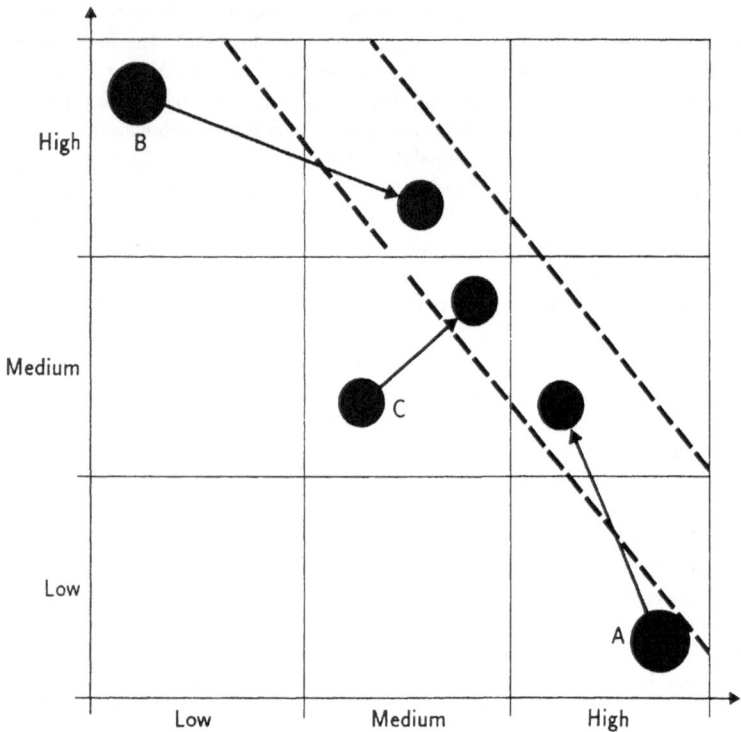

**Figure 4.12.** Capacity-flexibility problem of different strategies. Source: Ranta and Tchijov, 1990.

In the 1970s the potential value of multidimensional senses – especially *vision* and *touch* (*taction*) – began to be appreciated. The problem with "machine vision," as it is called, does not lie with the video camera itself but lies with the means of *interpreting* the picture.

Vision technology of the mid-1970s was *binary*. It detected and classified "blobs" based on their shapes, using statistical pattern recognition. The first generation of vision systems required a fairly powerful minicomputer, with specialized software to process visual information (pixels/sec) and discriminate patterns of shapes by "neighborhood." These early systems, largely

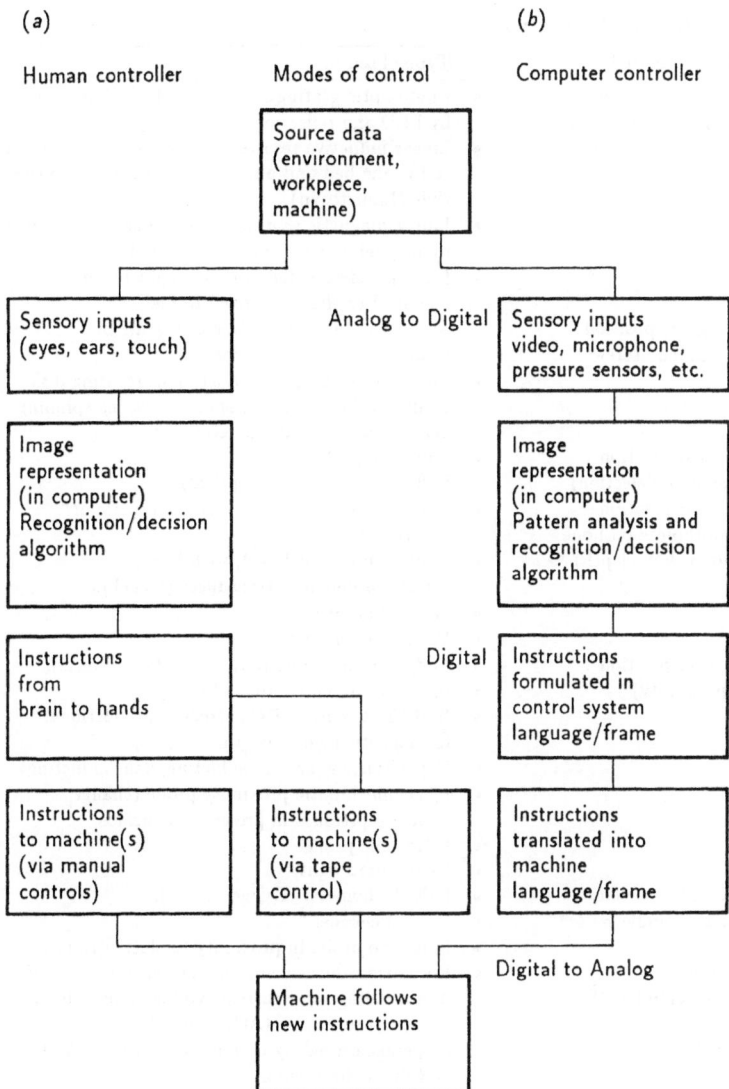

**Figure 4.13.** Modes of control

**Table 4.7.**  Sensor types.

| Data required | Type of sensor |
|---|---|
| Linear position (e.g., slide on bed) | • Photo-optic grating fixed to the bed illuminated by LED and reflections sensed by photodetector.<br>• Linear inductive resolver, i.e., consisting of a fixed coil in the bed and a moving coil mounted to the slide (Inductosyn).<br>• Linear magnetic grating fixed on bed and scanned by magnetic pickup on slide (+50 f).<br>• Laser interferometer (+12.5 f) split beam; one fixed to slide, one fixed to bed. |
| Angular position (e.g., lead screws) | • Angular industrial resolver consisting of rotating and fixed transformer.<br>• Photo-optic encoder, consisting of rotating disk of alternating clear and opaque sections spinning between an LED and a detector.[a] |
| Linear position (spindle defection) | • Inductive probe.<br>• Deflectometer (electro-optical). |
| Linear position (workpiece on bone diam. hole depth) | • Contact probe (caliper) with pneumatic or electric sensor.<br>• Contact probe (caliper), with linear variable differential transducer (LVDT).<br>• Magnetic grating.<br>• Laser interferometer. |
| Linear position (proximity) | • Eddy current probe (ferrous, surface objects).<br>• Inductive probe (ferrous objects).<br>• Hall effect, sensor; DC current is proportional to magnetic field strength.<br>• Capacitance gauges (conducting/nonconducting).<br>• Electropneumatic proximity probe (change in orifice pressure in presence of surface).<br>• Ultrasonic probe.<br>• "Structured light." |
| Angular velocity (e.g., spindle RPM) | • DEC tachometer (DC generator).<br>• Optical encoder.[a]<br>• Inductive probe in proximity to slotted ring. |
| Power (e.g., spindle HP) | • Watts transducer consisting of multiplication of motor voltage and current; voltage sensed by a voltage transformer (AC) or voltmeter (DC); amperage sensed by current transformer (AC) or voltage drop across a shunt in series with power; multiplication by IC chip. |

**Table 4.7.** Continued.

| Data required | Type of sensor |
|---|---|
| Angular torque (e.g., spindle torque or tool) | • Power (above) divided by spindle RPM (above) #3%. |
| | • Strain gauge dynamometer is most accurate. |
| | • Strain gauge dynamometer is most accurate to expensive method. |
| Linear acceleration (e.g., spindle vibration or deflection force) | • Piezoelectric accelerometer (quality crystal), voltage proportional to force applied. |
| Temperature | • Thermocouple (voltage proportional to temperature. |
| | • Thermistor (resistance inversely proportional to temperature). |
| Pressure (e.g., tactile force feedback from robot gripper or fingers) | • Silicon (MOS) strain gauge. |
| | • Piezoresistive transducer. |
| | • Conductive elastomer strain gauge. |
| | • Resistive potentiometer. |
| | • Inductive potentiometer. |
| | • Deflectometer (electro-optical). |

[a]Signal can be directly digitized.

experimental, were both crude and very slow. Machine vision systems became commercially available in the late 1970s, and a large number of new start-up ventures entered the field after 1980.

A second generation of vision systems, capable of discriminating *gray scales* and more sophisticated *syntactic* pattern recognition, became available to commercial users about 1982. Future systems will eventually add color, stereo, shading, texture, motion, shadows, and so on. However, it is not at all clear when these capabilities will appear in affordable commercial systems. Vision technology is currently "hot," and the apparent rate of technical progress is very high, as suggested by *Figure 4.14*. Nevertheless, adaptive systems employing sensory feedback – primarily vision or touch or both – are going to be the key to truly computer integrated "fifth generation" automation, as summarized in *Table 4.8*.

The key to improve performance of vision systems is parallel processing, and the key to reduce costs is "customized" VLSI chips. Such chips began to be produced in quantity by 1985. Tactile sensors will require parallel processing very similar to that needed for vision systems. It thus seems quite safe to project that adaptive control for both machine tools and robots

**Table 4.8.** Five generations of automation.

|  | Pre-manual control | First (1300) Fixed mechanical stored program (clockwork) | Second (1800) Variable sequence mechanical program (punched card/tape) |
|---|---|---|---|
| *Source of instructions for machine (How is message sent?)* | Human operator | Machine designer builder | Off-line programmer/ operator records sequences of instructions manually |
| *Mode of storage (How is message stored?)* | NA | Built-in (e.g., as patterns of cams, gears) | Serial: patterns as coded, holes in cards/tape |
| *Interface with controller (How is message received?)* | Mechanical: linkage to power source | Mechanical: self-controlled by direct mechanical links to drive shaft or power source | Mechanical: controlled by mechanical linkage actuated by cards via peg-in-hole mechanism |
| *Sensors providing feedback* | NA | NA | NA |
| *Communication with higher-level controller* | NA | NA | NA |

| Third (1950) Variable sequence electromechanical (analog/digital) | | Fourth (1975) Variable sequence digital (CNC) (computer control) | | Fifth (1990?) Adaptive intelligent AC          AI (systems integration) | |
|---|---|---|---|---|---|
| On-line operator "teaches" machine manually | Off-line programmer prepares instructions | Generated by computer, based on machine-level stored program instructions modified by feedback | | Generated by computer-based on high-level language instructions, modified by feedback | |
| Serial: as mechanical (analog) record (e.g., on wax vinyl disc) | Serial: as purely electrical impulses (e.g., on mag. tape) | In computer memory memory as program, with branching possibilities | | In computer memory as program with interpretive/adaptive capability | |
| Electromechanical: controlled by valves, switches, etc., that are activated by transducers – in turn, controlled playback of recording | | Electronic: reproduces motions computed by program, based on feedback info. | | Electronic: (as in CNC) adjusts to cumulative changes in state | |
| NA | | Narrow analog (converted to digital) (e.g., voltm./ strain gauge) | Spectrum digital (e.g., optical encoders) | Analog or digital, wide-spectrum, complete descriptions visual, tactile, requiring computer processing | |
| NA | | NA | Optional primary program down-loaded from higher level | Essential, i.e., micro-processor at machine level must pass visual and tactile info to higher levels to coordinate | High-level controller has ·learning ability |

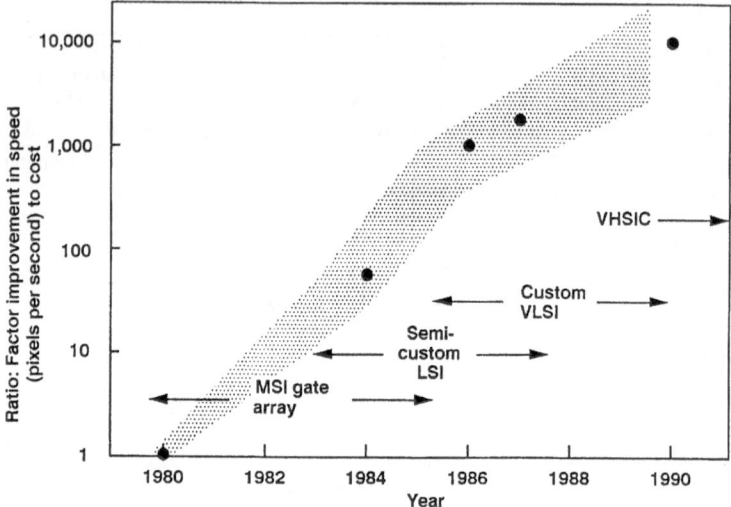

**Figure 4.14.** Estimated improvements in speed (pixels per sec)/cost ratio for neighborhood processing. Source: Funk, 1984.

using vision or tactile sensors or both will become a reality in the 1990s and will be fairly widespread by 2000, as shown in the last column of *Table 4.8*.

Current applications of vision systems are primarily for the control of manipulation tasks (such as drilling, routing, riveting, spot welding, soldering, sorting, palletizing, and assembly) and for inspection. In the case of inspection, the simplest use of machine vision is to check part dimensions against a stored template. Other types of inspection already exemplified include checking for integrity, color, orientation, reflectivity (shine), and so on. Automated inspection may become far more sophisticated in a few years, however, as judgment capabilities using artificial intelligence are built into the vision systems.

At present, most applications of vision (or taction) systems require substantial front-end investments in applications engineering. Moreover, they are still quite limited in their capabilities, primarily because of the difficulties in interpreting a visual scene. However, rapid technological improvements in the area of sensor sensitivity, software programmability, and user-friendliness, together with expected rapid cost reductions, will make entirely automated inspection a reality for most kinds of large-volume production by the year 2000 (if not sooner).

The control of complex processes obviously imposes substantial require- ments on the computer hardware, the machine interfaces, and the software. One limitation of nonadaptive numerical control, at present, is the fact that feed rate and cutting speed should be based on tool hardness and condi- tion (wear). An efficient and flexible programming system should allow for a range of each parameter. Obviously, the cumulative amount of wear depends on the initial tool hardness and on its cutting history. If the supervisory com- puter had in storage a good enough mathematical model for tool wear as a function of type of use, it would not need any sensory feedback from the tool itself. In turn, it could compute economically optimum feed rates and cutting speeds. However, this mode of (nonadaptive) control is inherently inflexible and intolerant of unexpected deviations.

In practice, too, the size of the required data base and the large compu- tational requirements of model-based controls seem to preclude doing these optimization calculations on-line in real time. A more practical approach might be to use semi-empirical models, e.g., the Taylor tool-life equations. But it is quite possible that the optimal condition as determined by such methods may violate physical conditions for safe (non-abusive) tool oper- ation (Yen and Wright, 1982). The problem is that the empirical tool-life equations only reflect physical wear mechanisms (such as abrasion, adhesion, cratering, fracture, and plastic deformation) in an average sense. Again, de- viations from the expected cause difficulty.

Yen and Wright propose an adaptive strategy. This requires a unified consideration of applicable physical constraints and economic optimization; the latter being carried out only after the former considerations are satis- fied. The applicable physical constraints are functions of dominant wear mechanism, but they can generally be expressed as localized cutting tool temperature and force limits. The adaptive control strategy, then, is to monitor these variables continuously, and adaptively vary the feed rates and cutting speeds to stay within the "safe" operating regime. This element of the overall control function can best be carried out at the individual ma- chine level, giving rise to the notion of distributed adaptive control. The higher-level supervisory computer needs only to be informed when a worn tool is actually replaced, so it can call for another one from stock or from the supplier.

The use of feedback information generated by sensors within the machine to provide decision information on the state of the machine or the state of the workpiece for the control computer requires further hardware–software interfaces. The sensor data must be "read" – usually as an analog signal – and converted into digital form. To be used in a decision algorithm, this signal must then be interpreted by comparing it with a stored or model-generated value or norm. For instance, if the actual metal removal rate is too low, as compared with the acceptable range of values, the MCU should inform the control computer and call for a tool change.

Most sensors currently used in industry produce low-grade, narrow-spectrum, yes–no signals. Such signals can only convey very simple messages and trigger correspondingly simple binary decisions. But there are many inspection situations where it is desirable to make complex adjustments or modifications in the instruction program to meet part specifications. For instance, if the force feedback from the work head increases beyond a certain point, the tool may be jammed. An "unjamming" routine must then be initiated. If it fails, the machine must then automatically stop and call for help from a supervisor. Or, if the workpiece may have arrived incorrectly oriented, sensory information must be adequate to enable the computer to interpret the situation correctly and issue instructions (to a robot arm, for instance) to reorient the part. Or, if a milling operation called for by the initial program results in a part with an incorrect physical dimensions, the computer must sense the situation and decide whether the flaw can be eliminated by additional milling. If so, the instruction program must be revised appropriately; if not, the part must be sent back for rework or discarded. In addition, the basic program should be permanently modified to ensure the problem is not repeated endlessly. Capabilities of this sort are at or beyond the state of the art because they require (1) sophisticated machine or tactile systems, (2) complex decision algorithms, and ultimately (3) an ability for the supervisory system to learn (i.e., reprogram itself) from experience.

It is clear, incidentally, that the capabilities described above must be a generalized part of the system's software, not specific to any machine or cell. This is so because the individual machines may be replaced or regrouped at any time – so, too, may the computers. The adjustment and modification capabilities must therefore be capable of interpreting sensory inputs that are also somewhat generalized in nature. This implies that the introduction of *artificial intelligence* (AI) into factory operation must be preceded or accompanied by the availability of machine vision and sophisticated wide-spectrum sensor (e.g., sensory information) processing capabilities. All of these capabilities evidently belong to the next (emerging) state of factory automation, denoted "fifth generation" automation in *Table 4.8*.

## 4.9  Artificial Intelligence

In the context of automation, the word intelligence has been overused, if not misused. The Japanese Industrial Robot Association (JIRA) classifies its robots into six categories, of which the most sophisticated is termed "intelligent," meaning merely that it is capable of being used adaptively in combination with sensors. In 1974 Quantum Science Corporation, a technological market-research firm, published a report called "The Intelligent Factory." It defined intelligence in a machine as "sufficient computing and memory capability to make decisions in response to sensory information from their environment." Examples of intelligent machines, by this criterion, included adaptive controls for machine tools, pattern recognition devices, assembly robots with sensory feedback, and inspection machines.

Other examples of confusion of sensory feedback and computational capabilities with true intelligence, in any meaningful sense, are common. For our purposes, therefore, it is important to be more precise. Sensory or computational facilities or combinations of both are not sufficient criteria for intelligence. Intelligence involves, at least, the ability to make (quasi) probabilistic inferences efficiently from incomplete information. At a higher level, it also involves the ability to learn – i.e., to improve its inferential skills – by referring to past experience.

In a manufacturing operation there are, of course, many potential applications of these abilities. One, already touched on, is "machine vision," i.e., the correct interpretation of two-dimensional pictures in terms of three-dimensional objects and how they are spatially related to each other. The importance of making inferences from incomplete data is clear in this case: in real situations nearby objects often block the view of those farther away. The interference problem can often be ameliorated by comparing several pictures from different viewpoints, but this also involves implicit adjustments for the different angles of view.

Tool-system failure diagnostics is another likely early application of artificial intelligence. Within the next decade, "expert systems" may become available to help production supervisors identify and solve operational problems that arise within the increasingly complex groupings of tools- and materials-handling devices. This sort of application is not too different, in principle, from some of the early successes in this field, such as Stanford's MYCIN (an expert system to advise physicians on drug therapy).

Another related, but more distant, application of artificial intelligence may be in the planning of cutting or forming sequences for three-dimensional parts, based on final shape specifications. While there are already some computer aided process planning (CAPP) models of this kind, they are generally specialized to a limited group of shapes. Moreover, they lack the ability to learn from experience, which a truly intelligent system should possess.

Forecasting trends is yet another possible example of applied intelligence. To be sure, the forecasting algorithm can be prespecified and fixed, in which case there is no intelligence involved except on the part of the programmer. However, an intelligent machine in the future might be given the ability to modify its own algorithms on the basis of comparing the results of its previous forecasts with subsequent reality. This sort of capability would be a significant addition to the current generation of MRP systems, for instance (see Chapter 6).

Natural language interfacing (both written and spoken) is a great and most elusive challenge for artificial intelligence. At present, a manager who wants to access data in a large data base must make his or her request through a specialist programmer, who can speak the "language" of the data base management system (DBMS). Obviously, if the DBMS were "intelligent," it could interpret a free-form natural language query (or ask questions) and eliminate the need for the specialist programmer at the interface. Such interfaces are being built, but progress has been painfully slow. For spoken language interfacing, the problem has been even more intractable, since the difficulties of interpreting ambiguous syntax are compounded by the difficulties of interpreting the sounds themselves. Yet, the advantages of developing such interfaces are potentially enormous and justify continuing investment in the area.

Artificial intelligence has engendered extraordinary optimism among some of its adherents. The so-called strong hypothesis "affirms that anything people can do, AI can do as well...[and] that artificial intelligence inevitably will equal and surpass human mental abilities – if not in twenty years, then surely in fifty" (Nilsson, 1984).

But high hopes for spectacular progress in this field have repeatedly been unfulfilled, leading to the suspicion that the problem is somehow much more difficult than the optimists have assumed. All in all, I cannot but agree with the findings of a US National Research Council (1983) report:

**Table 4.9.** Span of control.

| Span of computer control | Added to prior level | Software % of total investment |
|---|---|---|
| Stand-alone machine | Instructions for machine control | 2 |
| Machining center | Instructions for changing tools | 3 |
| Machining cell | Multiple-machine control | 4 |
| FMS (1) | Scheduling | 6 |
| FMS (2) | Loading/unloading, storage | 10 |
| FMS (3) | Inspection, sorting | 15 |
| Automated production line | Assembly, palletizing, kitting | 20 |
| Automated factory (1) | Computerization of functional modules, viz., MIS, MRP, CAD, CAPP, CAM | 40 |
| Automated factory (2) | Linkage of MIS, MRP, order processing, scheduling, cost analysis | 50 |
| Automated factory (3) | Linkage of CAD, CAE, CAPP, and CAM | 70 |

In an extremely narrow context, some expert systems outperform humans (e.g., MACSYMA), but certainly no machine exhibits the common sense facility of humans at this time. Machines cannot outperform humans in a general sense, and that may never be possible.

With regard to manufacturing, we anticipate modest and steady progress in all areas mentioned above, but no spectacular breakthroughs, at least within the next decade.

## 4.10 Summary

The evolution of CIM evidently proceeds in stages, characterized by increasing *span of control* (by computers) as shown in *Table 4.9*. As the span of control increases, of course, the relative importance of software and software engineering will become increasingly dominant (see *Figure 1.6* in Chapter 1). Only at a very high degree of computer integration – not yet technically feasible – will it be reasonable to speak of artificial intelligence in this area. A truly intelligent factory may well be built someday, but that day is still quite far in the future.

As a matter of some interest, it seems worthwhile to summarize the evolution of manufacturing control technology in terms of a specific case.

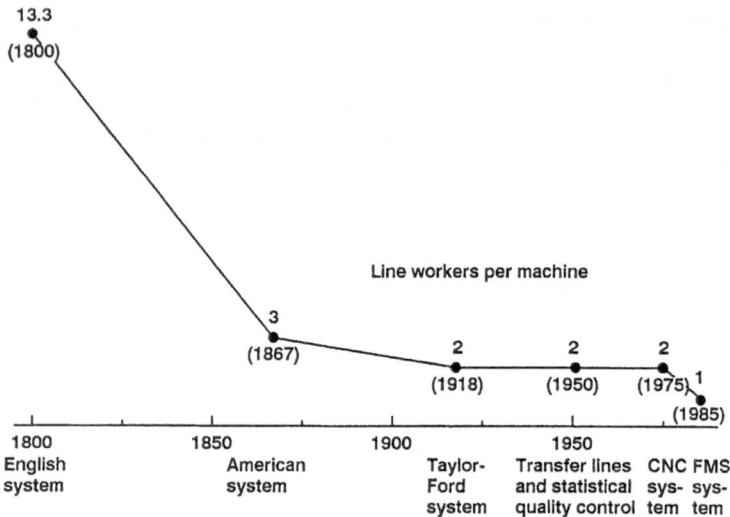

**Figure 4.15.** Comparison of six epochs in process control: Workers per machine. Source: Jaikumar, 1989a.

Inasmuch as I began this chapter with gun manufacturing, I may as well end with it. In Italy, near Venice, a family-owned gun manufacturer (Beretta) has been in business for close to 500 years (founded 1492). In 1800 the firm switched from a purely handicraft operation to what may be called the English system of manufacturing. Since then it has undergone five major reorganizations of its manufacturing operations (c. 1867, c. 1918, c. 1950, c. 1975, and c. 1985). The six "generations" are sketched in the following (Jaikumar, 1989a):

*English System* (c. 1800): 3 general purpose machines, 40 workers, 4:1 increase in labor productivity, 80% rework required, customized products ("infinite" variety), emphasis on individual skill and precision in manufacturing; instrument of control, micrometer.

*American System* (c. 1867): 50 specialized machines (plus jigs and fixtures), 150 workers (20 off-line), 3:1 increase in productivity, 50% rework, 3 different

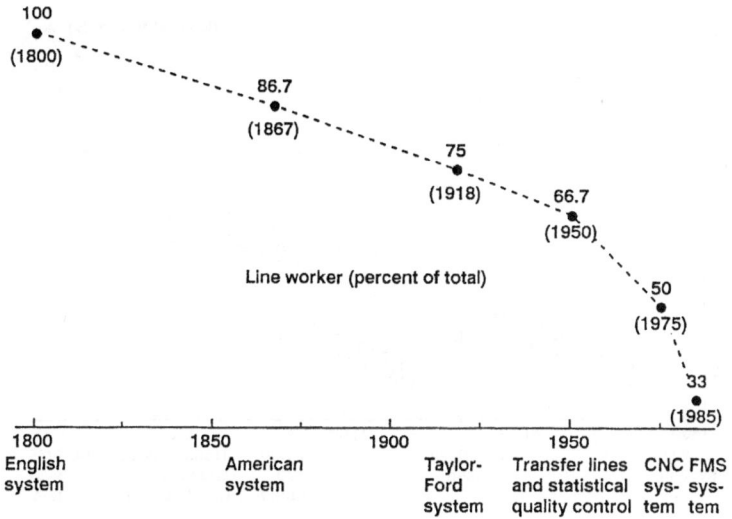

**Figure 4.16.** Comparison of six epochs in process control. Source: Jaikumar, 1989a.

models offered, emphasis on interchangeability of parts; instrument of control, go/no-go gauges.

*Taylor–Ford System* (c. 1918): 150 machines, 300 workers (60 off-line), 3:1 increase in productivity, 25% rework, 10 different models offered, emphasis on scientific management and specialization of function; instrument of control, stop watch.

*Statistical Quality Control (SQC) System* (c. 1950): 150 machines, 300 workers (100 off-line), 3:2 increase in productivity, 8% rework, 15 different models offered, emphasis on process control and information feedback from shop floor; instrument of control, control chart.

*Stand-Alone NC System* (c. 1975): 50 machines, 100 workers (50 off-line), 3:1 productivity increase, 2% rework, 100 different models offered, emphasis

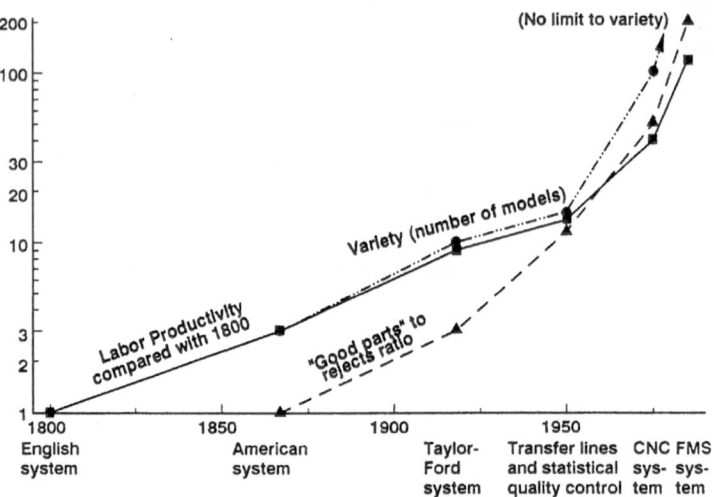

**Figure 4.17.** Comparison of six epochs in process control. Source: Jaikumar, 1989a.

on product-process integration, adaptability, and flexibility; instrument of control, electronic inspection.

*FMS System* (c. 1985): 30 machines, 30 workers (20 off-line), 3:1 productivity increase, 0.5% rework, ("infinite" product variety), emphasis on "process intelligence"; instrument of control, engineering work station.

It is tempting to believe that the Beretta experience may be an indicator of what can be expected in the future from other manufacturers. Gross output has increased enormously, of course. The overall increase in productivity since the pre-machine period is close to 500-fold. Yet absolute employment in manufacturing actually increased significantly until around 1950, when the first cuts occurred. Since then, direct employment in manufacturing has dropped by a factor of 10 (from 300 to 30), and the ratio of workers per machine has dropped from 13:3 to 1 (*Figure 4.15*). Meanwhile the ratio of off-line to on-line workers has grown continuously, from 13% in 1867 to 67% today (*Figure 4.16*). It is especially interesting that the early gains in productivity clearly owed a great deal to the design standardization that occurred between the English period and the American period. Recent gains in productivity (*Figure 4.17*) obviously owe nothing to standardization, since product diversity has been growing, but owe quite a bit to the dramatic quality improvements, as reflected in the drop in rework percentage from 80% in 1800 and 50% in 1867 to the present remarkable level of 0.5% – far below the level considered satisfactory by most companies today.

# Chapter 5

# Adoption/Diffusion of CIM Technologies

## 5.1 The Diffusion Process

The diffusion of a technology, such as CIM, may be regarded as a special case of a more general social diffusion process, viz., the diffusion of ideas, language, or life-style. One may also identify "similar" diffusion processes, such as the spread of an infectious disease or a pest. (In fact, the most commonly used diffusion models are known as "epidemic models.")

However, while these processes may offer some useful analogs, they also lack two key features of the process of technological diffusion. The first is the element of choice and decision. While people do *choose* whether to adopt a new concept, a new slang word, or a new hairstyle, much of the decision process is subconscious and is probably governed by factors such as the extent to which people want to behave *like* others, *unlike* others, or just do not care. Adopters have been classified into groups, ranging from *innovators* to *laggards* (Rogers, 1962). While not necessarily easy to determine in advance for a particular individual, the percentage of people in each category can probably be determined with reasonable accuracy for large populations, as suggested in *Figure 5.1*.

In the case of an infectious disease, of course, there is no conscious choice involved, except to the amount that people can control the extent of their exposure. If this is not controllable, the rate of spread is essentially deterministic.

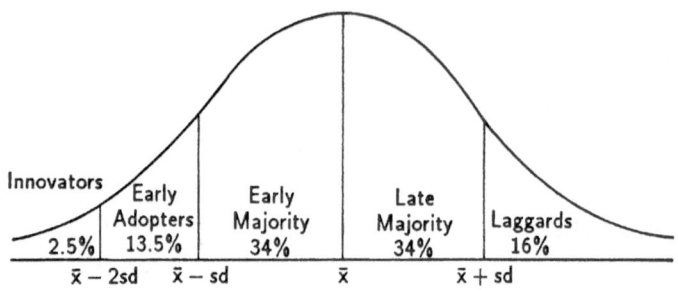

Adopter categorization on the basis of innovativeness

**Figure 5.1.** Adopter categorization on the basis of innovativeness.

By contrast, the adoption of a new product (other than an "impulse" item) does generally involve comparison, evaluation, and explicit choice. In most cases, there is not only a choice of brands or suppliers, but also a generic choice among alternative strategies or systems at several nested levels of abstraction. For instance, a factory manager may choose among robot vendors only after making prior choices among robot architectures (rectangular, cylindrical, spherical, revolute), drive systems (pneumatic, hydraulic, electric), and even between robots, ACVs, transfer lines, or human workers in the particular application.

This multilevel choice is always a factor in the adoption decision. Because it is so complex, the market success of a new product is extremely difficult to predict. For this reason, new consumer products are seldom introduced without extensive market studies – often involving distribution of the product in selected typical localities. Based on the results of such studies, changes in the product – or its packaging and presentation – are often made before full-scale introduction.

Unfortunately, new producer goods, or processes, cannot be pretested in this manner because the major cost of such a product is the development itself, not the distribution and marketing. Thus, even with the most careful attention to all controllable factors, R&D efforts sometimes result in failure, at least from an economic perspective. On occasion, the amounts of money involved have been extremely large, as in the case of Dupont's unsuccessful attempts to develop an artificial substitute for leather (Corfam) and an artificial substitute for silk (Qiana).

Once a new product has passed this initial hurdle, however, the uncertainties are sharply reduced. In effect, there is often a "yes/no" reaction from the marketplace that is quite hard to predict, at least on the basis of currently available methods. On the other hand, once the level of adoption has passed a certain point (a heuristic rule of thumb puts it at about 5% of the potential market), the likelihood of disappearance is quite low. The remaining issue is to estimate the subsequent trajectory of the process.

Abstracting from the complexity, economists have sought to explain the process in terms of a few simple explanatory variables. The first (and still most popular) econometric model assumes the diffusion follows a trajectory determined by the differential equation

$$\frac{df}{dt} = k(1 - f) \ , \tag{5.1}$$

where $f(t)$ is the fractional penetration of the market at time $t$ and $k$ is a constant. The solution of this equation is an S-shaped curve known as the logistic function. The first such model was used to explain the pattern of adoption of hybrid corn (Griliches, 1961). Later, the basic model was applied by Mansfield (1961) to explain technological diffusion in several industries. Mansfield's model can be explained using three terms in the equation

$$k = a + b\pi + cS + z \ , \tag{5.2}$$

where $\pi$ is proportional to the *profitability* of the innovation to the adopter, $S$ is proportional to the *size* or *magnitude* of the investment needed, and $z$ is an error term. The constants $a$, $b$, and $c$ are econometrically determined, but $a$ varies from industry to industry while $b$ and $c$ should be independent of sector. These constants were determined for 15 sectors by Blackman *et al.*, 1976.

Since the early 1970s a number of variations on this theme have been introduced by others. In general, they all use information gained from the early adoption history to parameterize the equations of the model (see Linstone and Sahal, 1976; Mahajan and Wind, 1986). More recent models tend to be more complex than the earlier ones, but the dependence of diffusion on profitability or *benefit* to the user is robust.

Unfortunately, most of the available diffusion models cannot be applied directly to a case like CIM (or computers), since they all assume – at least implicitly – that the "product" being diffused remains unchanged after the diffusion process starts. Yet, in the real world of computer technology there

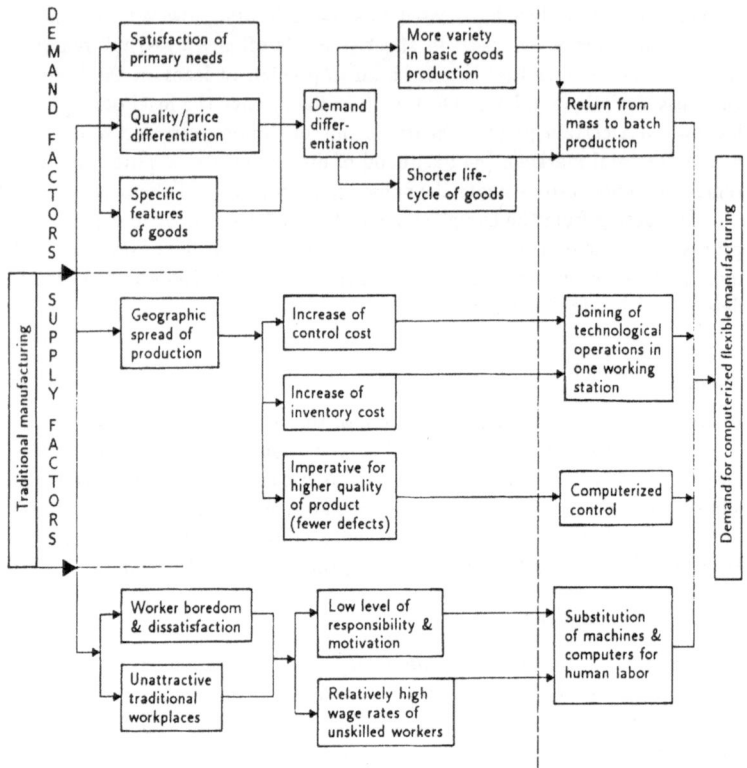

**Figure 5.2.** Factors driving the adoption of CIM.

is an active, conscious, feedback between the primary technology developer and the market, as suggested by *Figure 5.2.*

While controlled market testing is seldom possible, the early adopters of a new, evolving industrial technology (such as CIM) do provide a constant stream of useful information to the developers and manufacturers of producer goods, in terms of both the performance and reliability of the equipment and systems that have been put in service and the evermore clearly understood (and changing) needs of their customers.

A further complication is the fact that the evolving technology in the case of CIM (and other cases as well) arises from a variety of sources, not least of which are the early users themselves. Early users of NC machines, robots, and FMS systems, for instance, needed to learn how to use these devices effectively. In many cases, this depended on the development of special-purpose software, which remains proprietary (i.e., it does not diffuse to other potential users). It also depends on organizational learning, which is not easily transferable. Another important source of CIM technology, possibly the major one in this case, is the electronics industry, especially semiconductor and computer manufacturers.

Having made these points, it must be emphasized that the dominant mechanism governing diffusion is the calculus of *costs vs. benefits*. While technological information may not be equally available everywhere, lack of information is seldom the barrier to diffusion. In the major industrialized countries, at least, there cannot be significant differences in terms of basic technical information availability.

Nor does it make sense to assume that decision makers in one country are more or less rational than those in another. Any observed lags in diffusion between countries must, therefore, be explained by differences in the economic structure or the sociopolitical environment. Real labor costs; interest rates; energy costs; inventory ratios; domestic market size; labor, antitrust, and trade policy; currency stability; fiscal, monetary, and tax policy, and a host of other factors do differ somewhat from country to country and from sector to sector. With regard to the former, growth rates, interest rates (cost of capital), industrial policy, and labor laws are likely to be important; with regard to the latter, the importance of economies of scale and scope and the degree of standardization of the process are critical.

## 5.2 Progress vs. Diffusion

It is important not to confuse the processes of technological change/progress and technology diffusion/adoption. Although they often overlap to some degree, they are driven by different forces and controlled by different economic mechanisms. The adoption/diffusion process rarely occurs alone, as where a product or process is essentially fully developed before it reaches the marketplace. (An example of this might be a new drug such as penicillin or a chemical additive, such as tetraethyl lead.) A variant of this situation occurs when complex technologies are developed and introduced in distinguishable

**Table 5.1.** Labor equivalent of industrial robots.

|                        | 1 Shift |      | 2–3 Shifts |     |
|------------------------|---------|------|------------|-----|
| Source                 | B       | S    | B          | S   |
| Industrial robots, av. | 1.5     | 1.4  | 4.0        | 3.0 |
| Component manipulators | 2.1     | 1.6  | 6.2        | 3.3 |
| Tool manipulators      | 0.9     | 0.43 | 1.7        | 0.8 |

B = Battelle Frankfurt; S = Sociological Institute Göttingen. Source: Haustein and Maier, 1981.

"generations," as in the case of commercial jet aircraft or early computers. At the other extreme is a technology, like the telephone or the electric light, that is introduced in a relatively primitive form and continuously improved thereafter. Here progress and diffusion occur in parallel, and interact closely.

Computer integrated manufacturing (CIM) is an example of the latter type. In fact, the term "diffusion" may even be misleading, since it seems to imply a single center where CIM technology is created, and from which it spreads out spatially in all directions. This image is clearly wrong, inasmuch as there are many such centers, in scores of institutions located on several continents. The most that can be said in terms of identifying "leaders," and (by implication) "followers," is that CIM is an outgrowth of electronics technology, although leadership in the parent technology does not necessarily coincide, or collocate, with leadership in downstream applications technology. This is an unusual pattern, in historical terms.

## 5.3   Benefits Measurement

The various hypotheses stated in Chapters 1–4 imply that improved product quality and increased flexibility in the use of capital are beneficial to users of CIM. However, the argument thus far is only qualitative. To carry it a step further, one must define quality and flexibility more precisely and formulate them in terms of conventional economic variables and models. This is the next task to be undertaken, and it is a vital one.

To organize the discussion, it is helpful to consider five possible kinds of economic benefit:

(1) *Labor saving.* Some CIM technologies (most notably robots) can be regarded as direct substitutes for semiskilled human labor, as indicated in *Table 5.1* and *Figure 5.3*. This means that robots (sometimes called "steel-collar workers") can also be regarded as additions to the labor

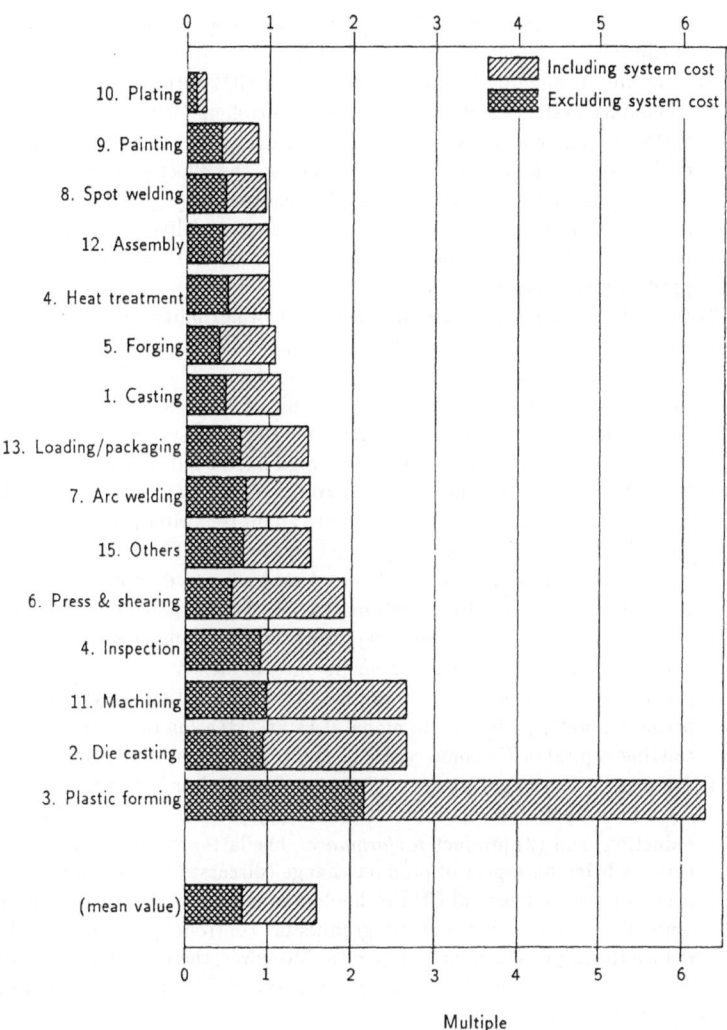

**Figure 5.3.** Japanese managers' willingness to pay for robots, expressed as multiples of the annual cost of a worker.

force, although their "wages" are partly operating costs and partly costs of capital.

(2) *Capacity augmenting/capital saving.* Some CIM technologies, such as scheduling systems and programmable controllers (PCs) with sensory feedback, can be regarded as capital savers or capacity augmenters. This is the case to the extent that they increase the effective utilization of existing machine tools and other capital equipment (e.g., by permitting unmanned operation at night) or permit shorter delivery times, faster turnarounds and reductions in the inventory of work in progress. The productivity of capital is thus increased.

(3) *Capital sharing/saving.* The major benefit of flexibility, as discussed in Chapter 3, is that it permits faster response to changing market conditions, or superior ability to differentiate products. The major reason for slow response is the widespread use of dedicated, specialized ("Detroit") automation in mass production. Here, the lowest possible marginal unit cost is achieved at the expense of very high fixed capital investment and large write-offs in case the product becomes obsolete and cannot be sold. Flexibility in this context is the ability to adapt (or switch) capital equipment from one generation of a product to the next. The term flexibility is also widely used in a rather different context – to describe a futuristic concept analogous to an automated job shop, capable of producing "parts on demand." In either case, capital is shared among several products rather than dedicated to a single one. Evidently capital sharing is practically indistinguishable from capacity augmentation. However, it is perhaps slightly preferable to model it as an extension of the lifetime of existing capital or (in some cases) as credit for capital recovery.

(4) *Product quality improvement.* The term "quality" is not very precise, since it comprises at least two aspects: (1) product *reliability* (defect reduction) and (2) product *performance.* The latter can be disregarded, here, as being an aspect of product change (discussed in Section 5.4). It is postulated that several CIM technologies, especially the use of "smart sensors" in conjunction with programmable controllers, will eventually reduce the in-process error/defect rate. Moreover, these technologies will also permit more complete and more accurate testing and inspection of workpieces and final products.

(5) *Acceleration of product performance improvement.* As noted above in connection with quality, improved product performance can be distinguished, in principle, from improved product reliability though reduced error/defect rates. The latter is a function of the manufacturing process

**Table 5.2(a).** Benefits of FMS (Case No. 1).

|  | Before | After |
|---|---|---|
| Types of parts/month | 543 | 543 |
| Number of pieces/month | 11,120 | 11,120 |
| Floor space | 16,500 m$^2$ | 6,600 m$^2$ |
| CNC machine tools | 66 | 38 |
| Non-NC machine tools | 24 | 5 |
| Total machine tools | 90 | 43 |
| Operators | 170 | 36 |
| Distribution and production control workers | 25 | 3 |
| Machining time per part[a] (days) | 35 | 3 |
| Unit assembly | 14 | 7 |
| Final assembly | 42 | 20 |
| Total time | 91 | 30 |

[a]Including queuing (work in progress). Source: Jaikumar, 1989b.

**Table 5.2(b).** Benefits of FMS (Case No. 2).

| Auto engines | Before | After |
|---|---|---|
| Numer of engines/month | 5,000 | 5,000 |
| Floor space | 6,000 m$^2$ | 10,000 m$^2$ |
| Machine availability | ? | 95% |
| Capital cost | 100% | 60% |
| Work force | ? | 50 |
| Lead time |  | no change |
| WIP |  | no change |

Source: Merchant, 1989.

**Table 5.2(c).** Benefits of FMS (Case No. 3).

| Sewing machines | Before | After |
|---|---|---|
| Number of products/month | 8,000 | 8,000 |
| Number of different models | 30 | 30 |
| Number of different parts | 60 | 60 |
| Work force per shift | 28 | 3 |
| Minimum lot size | 150–200 | 1 |
| Changeover time | 3 hours | 1 minute |
| Finished parts inventory | 6 months | 2 days |

Source: Merchant, 1989b.

**Table 5.2(d).** Benefits of FMS (Case No. 4).

| Machine tools manufacturing | Before | After |
|---|---|---|
| Floor space (m$^2$) | 6,500 | 3,000 |
| No. of machines | 68 | 18 |
| Staff, total | 215 | 12 |
| Inventory ($ mill.) | 5.0 | 0.218 |
| System cost ($ mill.) | 14 | 18 |
| Throughput time (days) | 90 | 3 |

Source: Fleissner, 1987, p. 101.

**Table 5.3.** Average FMS benefits by area of application (percentage).

| Indicators | Metal cutting | Metal forming | Welding & assembly | Overall average |
|---|---|---|---|---|
| Pay-back time (years) | 3.8 | 3.1 | 3.6 | 3.8 |
| Lead time reduction (%) | 80.39 | 89.47 | 90.10 | 81.48 |
| In-process time reduction (%) | 86.30 | 75.00 | 75.61 | 83.05 |
| Inventory reduction (%) | 74.36 | * | * | 76.19 |
| Work-in-progress reduction (%) | 72.97 | * | * | 75.00 |
| Personnel reduction (%) | 76.19 | 44.44 | 75.61 | 74.36 |
| Number of machines reduction (%) | 75.00 | * | * | 75.61 |
| Floor space reduction (%) | 67.74 | 41.18 | 54.55 | 64.29 |
| Capacity utilization increase (%) | 80.00 | * | * | 80.00 |
| Unit cost reduction (%) | 41.18 | 62.96 | 58.33 | 44.44 |

*Number of observations is not enough for averaging. Source: IIASA.

only, whereas the former requires changes in the actual design of the product. It was pointed out that one benefit of flexibility is that it reduces the cost of each product change. A further benefit is that, as a result, product redesigns are likely to be more frequent. The problem for an economist, is to find empirical evidence of a relationship between the cost of product redesign and retooling and the rate of product performance improvement. This appears to be a relatively unplowed field of research, to date.

The "mix" of benefits varies considerably from application to application. *Tables 5.2(a–d)* taken from the literature and *Table 5.3* from the IIASA data base, are illustrative.

# 5.4 Static vs. Dynamic Approaches to Benefits Measurement

Up to this point, we have not focused on the question: benefits to whom? In fact, this is a critical issue. In a free-enterprise capitalist system, short-run benefits are likely to be appropriated mainly by producers (as profits), which would in the first instance most likely be reinvested. In the long run, in a competitive economy, essentially all benefits will be passed on to consumers through product price reductions, performance improvements, and wage increases.[1] In a unionized industry or a socialist economy, short-run benefits (if any) are more likely to be appropriated directly to workers, through higher pay, shorter hours, or costly "job security" measures, or through all of these measures.

More important for our purposes, it is *only* the short-term benefits appropriable as profit by producers that can directly motivate innovation and technological diffusion (Mansfield, 1961, 1968). In this context, it is clear that in a static environment, labor saving, capacity augmentation, and capital sharing may contribute immediately to profitability. On the other hand, product quality and performance improvements may have a less direct impact on profitability in the short run, except to the extent that error/defect control has a direct effect on costs.

In a static world of competitive *price-takers* and given *fixed* and *variable* costs, the optimum (short-run profit maximizing) protection level is determined by the shape of the variable cost curve. Assuming the usual U-shaped variable cost curve (Chapter 3), the optimum production level is found by equating marginal revenue and marginal cost. If demand increases, but total capacity remains fixed, prices and profits will rise and vice versa. If the industry is profitable, any existing (or new) producer can increase capacity, thus reducing the average costs and break-even point; the producer can increase market share by price cutting. But if several producers do this, the result is overcapacity and losses. Moreover, assuming nonconvertible (inflexible) capital, the effective marginal cost now becomes the marginal *variable* cost, and each competitor will go on producing even if it earns no net return on capital. In fact, the static competitive market is inherently unstable (and therefore not static) at any finite profit level.

In other words (as Joseph Schumpeter pointed out long ago), profits in a competitive market are inherently a dynamic phenomenon reflecting an exploitable temporary cost or price advantage. The advantage at any moment in time may be due to superior brand-name recognition, cheaper labor or energy source, better location vis-à-vis markets, more efficient production technology, or better product design. But unless these advantages are protected, e.g., by brand-name copyright (e.g., "Coke"), a monopoly franchise (such as CBS), an impenetrable secret, or a set of interlocking patents, profitability will last just as long as it takes for a competitor to imitate or improve on the product or build a larger or newer plant or both.

It follows, therefore, that continuous long-term profitability for a firm can only be assured by a *continuous* process of creating and exploiting new advantages (of some kind) to replace the older, dissipating ones. Opening new markets, advertising, product improvement, process improvement – all are means of creating competitive advantages. Forward motion is essential: to be inactive is to sink and be overwhelmed. A moving bicycle, a circling "hula-hoop," a spinning top, or a gliding water ski are dynamically stable; but when the motion stops the system collapses. The same thing holds true for species in an ecosystem or a firm in a competitive market. There is no safe place to hide indefinitely from hungry predators seeking a meal or hungry competitors seeking a market.

In short, only a dynamic model of firm behavior has any value in assessing the benefits of CIM technologies (or, indeed, any other technologies provided exogenously). Furthermore, it is essential to view the firm in the context of its dynamic competitive environment. Most simple models of the behavior of the firm assume a static environment (e.g., an exogenous demand schedule or market price) and neglect the realities of competitive response. If all competing firms adopted a more efficient production technology simultaneously, none would gain any special advantage over the others but all would bear the cost of the necessary investment. To the extent that the adoption of more efficient product processes (CIM) results in lower costs and these are subsequently passed on to consumers as lower prices, the market for each product might (or might not) grow enough to result in increased profitability for the producer, *ceteris paribus*.

## 5.5   Numerical Control (NC and CNC)

Adoption of the first generation NC machines was slow. By 1963 only about 2,000 NC machines were in service in the USA and very few in the rest of the world. One reason was the high cost of controllers. An early (1958) transistorized control unit cost $70,000 to $80,000. By 1968 this had fallen

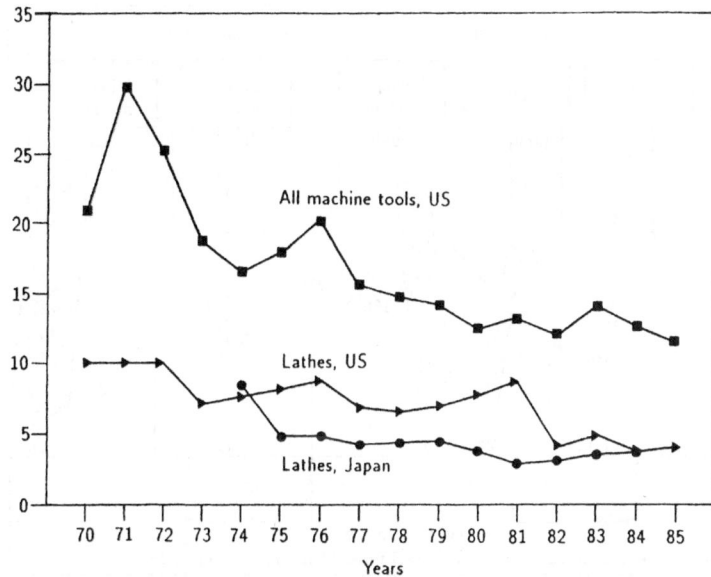

**Figure 5.4.** Cost of NC machine tools, expressed as multiples of the average cost of non-NC machine tools.

to $30,000. An improved controller employing integrated circuitry (c. 1974) cost $15,000. Application of LSI technology in the early 1970s brought the costs down even faster while simultaneously providing for vastly increased capability.

Despite this, NC and CNC machine tools still cost far more than comparable machine tools that are manually controlled. The ratio of actual purchase cost per machine tool averaged over the entire US market (shown in *Figure 5.4*) ranged between 20 × and 30 × prior to 1973, and this ratio is still of the order of 10 ×. Part of the explanation of this enormous price difference is that machining centers (which are very costly, and mostly CNC) are being averaged together with simple shop tools such as drills which are not numerically controlled. The turning machine (lathe) category does not include machining centers, and offers a more realistic comparison. But even in this case, the price ratio was about 10 × in the early 1970s and has declined to around 4 × today. Note that the ratio was considerably higher in

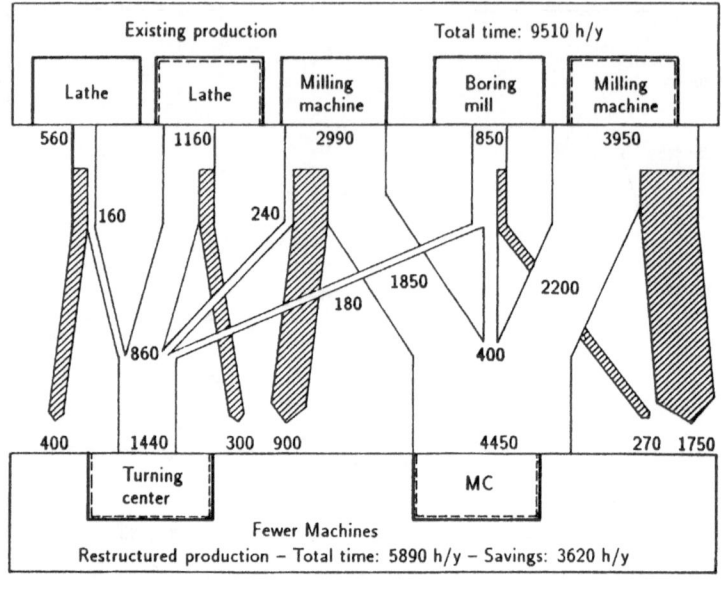

**Figure 5.5.**   Benefits of tool changing capability (machining centers). Source: Warnecke, 1989.

the USA than in Japan, prior to 1982, when Japanese imports forced US prices down.

The high prices of NC machines are compensated for by high utilization rates. Whereas a conventional stand-alone machine tool is productively cutting only 3% to 10% of the time, an NC or CNC machine can be productive 50% or more of the time (Cook, 1975). A rule of thumb for NC application is that productive output per machine rises by 3 × while unit costs (taking into account lower labor requirements) fall by one-third or so. A recent trend has been for one or two numerically controlled (CNC) machining centers to replace a number of more specialized machines, either manual or NC. An example of this is shown in *Figure 5.5*; a single turning center and

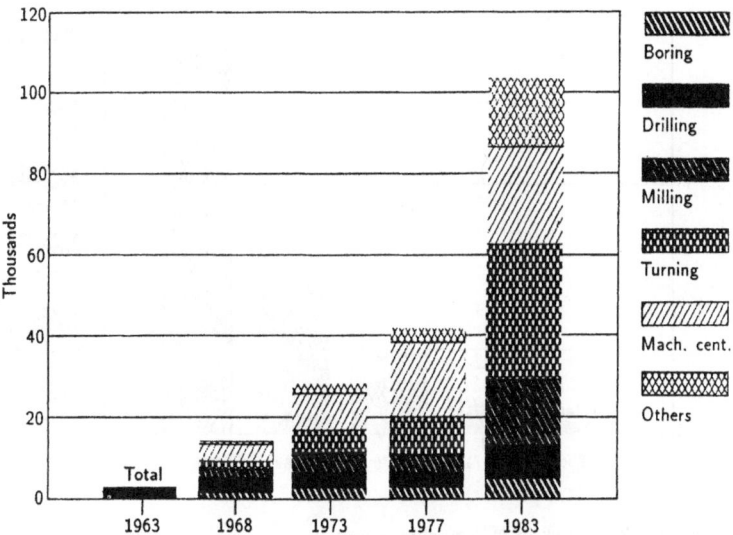

**Figure 5.6.** Stock of NC machine tools by type, USA.

a single machining center (MC) replace five stand-alone machines (including an NC lathe and an NC milling machine) with a 38% overall savings in machine-hours per year. In the reorganized system the CNC machining center operates at nearly 51% of theoretical capacity (8,760 hours per year) as compared with 45% for its predecessor, the NC milling machine, and 34% for the manually controlled milling machine. Utilization rates for turning machines tend to be lower, but the increases are comparable.

In the early 1970s expectations for the penetration of NC and CNC were overoptimistic, because high prices were difficult to justify by measured benefits. The reality was a much more modest (though still noteworthy) growth in the use of NC/CNC. Still, by 1983, NC and CNC machines accounted for one-third of all new machine-tool purchases in the USA, and more than 103,000 NC and CNC machines were in service. By 1989 the number of NC and CNC machines in the USA had more than doubled to 221,000 units. *Figures 5.6* and *5.7* show the growth in usage by type in the USA and the UK. Although this category now represents only about 10% of all machine

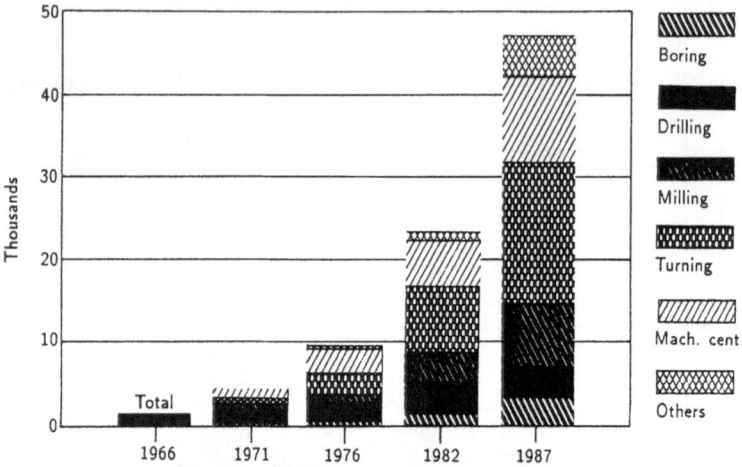

**Figure 5.7.** Stock of NC machine tools by type, UK.

tools in the USA, it accounts for a much higher (but not accurately known) percent of metal-working output.

Bearing in mind that many machine tools are not used for production, and that many production machines are specialized and automatic, it is likely that NC/CNC has already achieved around 25% penetration of its maximum present potential. However, whereas most NC and CNC machine tools were used in a "stand-alone" mode prior to the 1980s, more and more of them are being incorporated into flexible manufacturing cells (FMCs) and systems (FMSs). This, in turn, increases the range of potential applications, as shown schematically in *Figure 5.8*. In effect, CNC is encroaching on the (former) domain of manual production, on one side, and on the former domain of mass production, on the other.

## 5.6 Robots

Robots were introduced commercially in 1959. Initial acceptance was very slow. Only some 200 industrial robots were in service in the USA by 1970. The first Japanese robot appeared in 1969 (Kawasaki, a licensee of Unima-

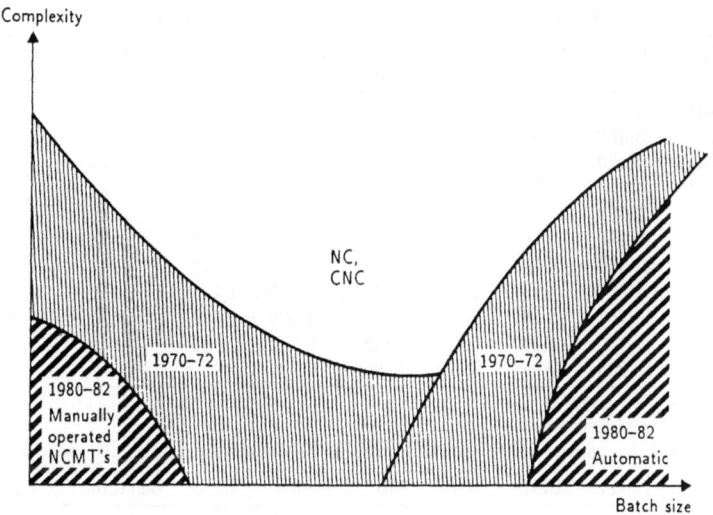

**Figure 5.8.** Increasing range of application, NC machines. Source: Ray, in Nabseth and Ray.

tion). Demand surged in Japan soon after, and picked up somewhat in the USA and elsewhere in the early 1970s. By 1974, when CNC capabilities became available, there were about 1,100 robots in service in the USA, and expectations exploded. (An optimistic 1975 market report anticipated that 24,000 robots would be in service by 1977.) Japan has continued to lead in robot use (and production), as shown in *Table 5.4*. At the end of 1989, Japan had 175,000 robots in service (68% of the world's total) as compared with 33,000 in the USA. Since robots are, essentially, substitute workers, the ratio of robots to manufacturing workers is a good measure of penetration (*Table 5.5*).

The exceptionally rapid introduction in Japan seems to be explained by three factors. The first was demographic: unlike the USA and Europe (except Germany), Japan had a severe labor shortage in the 1970s and robots were regarded as a labor-substituting technology (Mori, 1989). *Table 5.1* (labor substitution ratio) and *Figure 5.3* (Japanese entrepreneurs' willingness to pay to replace a worker) tend to confirm this statement. Further confir-

**Table 5.4.** Industrial robot population in selected countries.

| Year | Japan | USA | UK | FRG | France | Italy | Belgium | Sweden |
|------|-------|-----|-----|------|--------|-------|---------|--------|
| 1974 | 1,000 | 1,200 | 50 | 130 | 30 | 90 | | 85 |
| 1975 | 1,400 | | | | | | | |
| 1976 | 3,600 | 2,000 | | | | | | |
| 1977 | 4,900 | | 80 | 541 | | | 12 | |
| 1978 | 6,500 | 2,500 | 125 | | | 300 | 21 | 415 |
| 1979 | 9,100 | | | | | | 30 | |
| 1980 | 14,250 | 3,400 | 371 | 1,255 | 580 | 454 | 58 | 795 |
| 1981 | 21,000 | 4,700 | 713 | 2,300 | 790 | 691 | 242 | 950 |
| 1982 | 31,857 | 6,250 | 1,152 | 3,500 | 1,385 | 1,143 | 361 | 1,400 |
| 1983 | 46,757 | 9,387 | 1,753 | 4,800 | 1,920 | 1,850 | 514 | 1,600 |
| 1984 | 67,300 | 14,550 | 2,623 | 6,600 | 2,750 | 2,585 | 860 | 1,900 |
| 1985 | 93,000 | 20,000 | 3,017 | 8,800 | | | | |

Source: Tani, 1989.

**Table 5.5.** Industrial robot (IR) population density (per thousand workers).

| Year | Japan | USA | UK | FRG | France | Italy | Belgium | Sweden |
|------|-------|-----|-----|------|--------|-------|---------|--------|
| 1974 | 0.083 | 0.059 | 0.006 | 0.014 | 0.005 | 0.017 | 0.000 | 0.127 |
| 1975 | 0.123 | | | | | | 0.000 | 0.000 |
| 1976 | 0.318 | 0.105 | | | | | 0.000 | 0.000 |
| 1977 | 0.435 | | 0.011 | 0.065 | | | 0.013 | 0.000 |
| 1978 | 0.586 | 0.122 | 0.017 | | | 0.064 | 0.023 | 0.683 |
| 1979 | 0.822 | | | | | 0.000 | 0.034 | |
| 1980 | 1.256 | 0.168 | 0.053 | 0.149 | 0.111 | 0.096 | 0.067 | 1.308 |
| 1981 | 1.823 | 0.233 | 0.115 | 0.281 | 0.156 | 0.149 | 0.294 | 1.578 |
| 1982 | 2.768 | 0.333 | 0.196 | 0.442 | 0.277 | 0.252 | 0.456 | 2.418 |
| 1983 | 3.979 | 0.509 | 0.313 | 0.631 | 0.393 | 0.420 | 0.665 | 2.920 |
| 1984 | 5.553 | 0.751 | 0.476 | 0.878 | 0.580 | 0.615 | 1.126 | 3.565 |
| 1985 | 7.530 | 1.036 | 0.548 | 1.159 | | | | |

Source: Tani, 1989.

mation is the significant correlation between average hourly wage rates and robot usage, found by Tani (1989).[2]

Second, manufacturing output was growing much faster in Japan than in the other countries. It is inherently much easier to find useful tasks for robots in newly designed factories where they can work with other machines than it is to embed stand-alone robots retroactively in older plants.

Third, in Japan early users often built their own robots and later sold them for specific applications. The first generation of commercial robots

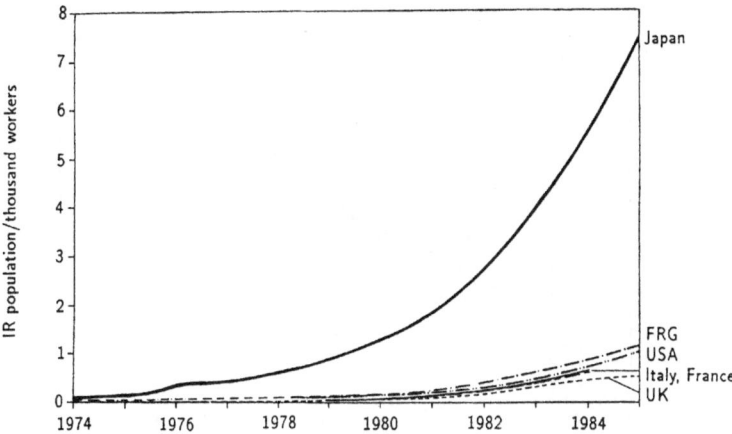

**Figure 5.9.** Industrial robots (IRs) per thousand workers.

in the USA were designed without specific application in mind, and both producers and manufacturers underestimated the cost of applications engineering. Programming languages for robots are diverse and still relatively clumsy. Thus engineering costs for new applications tend to be quite high (up to double the cost of the robot itself), which is a major impediment to small-volume and first-time users (Miller, 1983). Moreover, first generation US robots used analog rather than digital controls, and were therefore suitable *only* for stand-alone applications. They were inherently unsuitable for linking with machine tools under common computer control. The Japanese, starting later, successfully avoided this trap.

It is noteworthy that most major industrial countries have adopted robots at about the same *rate*, but with different characteristic *lag times* (*Figure 5.9*). The best fit projection (*Figure 5.10*) corresponds to lags of 4.3 years (USA), 4.9 years (FRG), 5.8 years (Italy), 6.3 years (France), and 7.5 years (UK).

One notable aspect of international robot diffusion is the unusual concentration of assembly robots in Japan (*Table 5.6*). This fact is sometimes cited as evidence of greater aggressiveness in pursuing advanced automation, and to some degree this is true. However, there is another partial explanation of the phenomenon. It happens that assembly robots, to date, are mainly suited for putting together rather small, lightweight items made in many

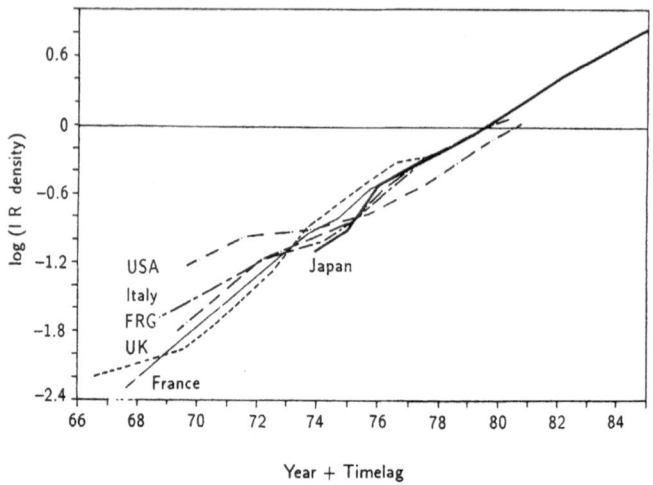

Year + Timelag

**Figure 5.10.** Growth of IR use on log scale, adjusted for lags. Source: Tchijov, forthcoming.

different models and variants. Products suitable for robot assembly include printers, cameras, watches, and consumer electronics. All of these are industries now dominated by Japan. Thus it is not surprising, and certainly not an indication of "lag" on the part of other countries, that Japan has far more assembly robots than other industrial countries, a very large percentage of which (40%) are involved in assembly tasks. By contrast, the UK and West Germany have, respectively, 9.7% and 8.6% of their robots doing assembly tasks (Tani, 1989).

It is very difficult to estimate the maximum level of penetration of robots. In the case of robots, a simplistic calculation based on the substitution of one robot for every two workers in the semiskilled machine operative category (excluding transport operatives) suggests an ultimate potential of three to four million robots in the US manufacturing sector alone and twelve to fifteen million manufacturing robots worldwide. This is much too high a number, if the potential for 24 hours a day operation is realized. On the other hand, robots will not replace all operatives – especially in smaller firms – for at least two to three more decades. Any such massive replacement presupposes dramatic improvements in robot programmability and performance and price

**Table 5.6.** Application distribution of IR.

| Application | Japan 1982–1985 % | UK 1985E* % | FRG 1985E* % | Italy 1984E* % | Belgium 1984E* % | Spain 1985E* % |
|---|---|---|---|---|---|---|
| Welding (Spot) | 9.2 | 16.9 | 29.0 | 28.0 | 60.0 | 50.2 |
| Welding (Arc) | 13.9 | 13.6 | 20.2 | 10.8 | 7.3 | 13.3 |
| Assembly | 39.9 | 9.7 | 8.6 | 11.8 | 0.5 | 6.4 |
| Loading/Unloading | 6.3 | 9.5 | 9.2 | 26.5 | 8.4 | 15.4 |
| Painting | 2.2 | 6.4 | 8.8 | 8.9 | | 6.8 |
| Injection moulding | 13.9 | 18.3 | | | | |
| Inspection/Test | 1.2 | 1.9 | | 1.2 | | 2.1 |
| Others | 13.9 | 23.7 | 24.2 | 12.8 | 23.8 | 5.8 |
| (Educational, etc.) | | (5.5) | (2.4) | | (11.4) | |

*E means at the end of the year. Source: Tani, 1989.

cuts. Moreover, the number of replaceable operatives is being reduced by other factors, such as the introduction of FMS (Section 5.7).

All things considered, the present level of penetration of robots is quite low in relation to maximum "technical" potential based on current industrial structure and employment patterns (Ayres and Miller, 1983). According to one estimate by a West German auto manufacturer, in the early 1980s, robots had already replaced 9% of the potential jobs, with an additional 7% substitution possible based on current technology and 10% for next generation robot technology. A further 20% was thought to be "conceivable," leaving 54% of the possible jobs beyond the reach of robots (Ebel, 1985). Based on this analysis, robots (c. 1982) had reached around 20% of their potential, and current figures would presumably be in the 50% range for that auto manufacturer.

Needless to say, auto manufacturers have been in the vanguard of robot users, and the penetration level for the metal-working sector as a whole is undoubtedly much lower. On the other hand, there are already some indications that most of the potential efficiency benefits to be gained by substituting robots for humans have already been achieved, at least in Japan. Based on econometric analysis, the average robot "replaced" from two to five workers between 1970 and 1973, as shown in *Table 5.7*. However, since 1977 the average ratio lies between 1:1 and 1:3 workers per robot added, and the ratio is essentially constant. Further substitution now appears to be driven mainly by wage increases, not efficiency gains.

**Table 5.7.** Price index (1980 = 1), average labor cost per worker (million yen), and equivalent workers per robot.

| Year | Price index | Labor cost per worker | Equivalent workers per unit (low) | Equivalent workers per unit (high) |
|------|-------------|-----------------------|-----------------------------------|------------------------------------|
| 1970 | 1.77857  | 0.95474 | 4.305 | 5.017 |
| 1971 | 1.19354  | 1.10263 | 3.051 | 3.561 |
| 1972 | 1.69316  | 1.25622 | 2.542 | 2.968 |
| 1973 | 1.75534  | 1.45976 | 2.210 | 2.579 |
| 1974 | 1.15822  | 1.95296 | 1.887 | 2.201 |
| 1975 | 1.31297  | 2.3468  | 1.582 | 1.846 |
| 1976 | 0.88558  | 2.57803 | 1.355 | 1.582 |
| 1977 | 0.865553 | 2.88791 | 1.248 | 1.455 |
| 1978 | 0.982675 | 3.1426  | 1.181 | 1.378 |
| 1979 | 0.945306 | 3.31669 | 1.126 | 1.314 |
| 1980 | 1.0      | 3.5441  | 1.100 | 1.283 |
| 1981 | 0.944806 | 3.74214 | 1.093 | 1.275 |
| 1982 | 0.838704 | 3.95995 | 1.096 | 1.279 |
| 1983 | 0.768663 | 4.06842 | 1.111 | 1.297 |
| 1984 | 0.704833 | 4.26936 | 1.120 | 1.308 |
| 1985 | 0.680084 | 4.39102 | 1.107 | 1.292 |

Source: Mori, 1989.

Increasingly, robots will be integral components of FMSs and CIM facilities, rather than stand-alone units. Thus, the future adoption of robots will be determined by the adoption of other CIM technologies.

## 5.7   Flexible Manufacturing Cells and Systems

As noted above, "islands" of CNC machine tools are being linked together into flexible cells (FMCs and FMSs). The first few such systems were conceived of and built in the late 1960s. By 1978 the world population of FMSs was around 100. Since then, growth has been explosive – ten-fold in a decade (*Figure 5.11*). Costs have declined significantly since the early 1970s, but only by a factor of 2 (*Figure 5.12*) – not enough to explain the extraordinary growth rate in FMS adoption. The increasing demand for flexibility and diversity in (formerly) mass-production-oriented industries is apparently the key.

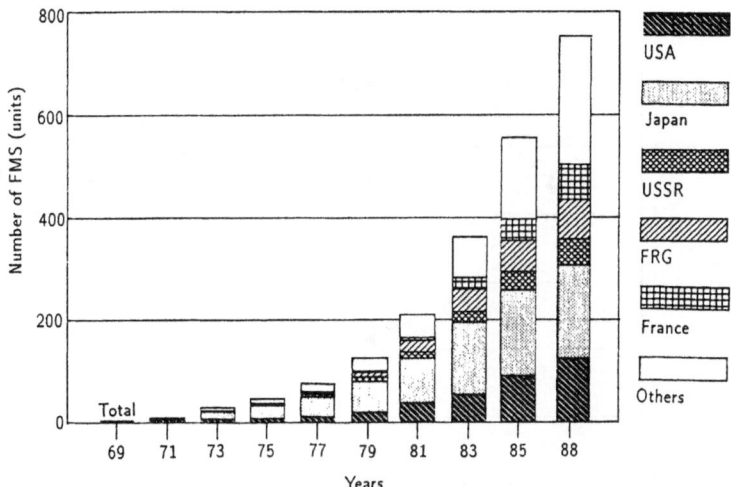

**Figure 5.11.** Use of FMSs, by country. Source: Tchijov, forthcoming.

As in the case of NC/CNC machine tools, there is evidence that the *domain* of FMC/FMS is growing. Most early units in the USA were designed, effectively, as alternatives to mass-production systems based on transfer lines, for products made in a small number of sizes or variants and relatively large batches. (Caterpillar was a typical pioneer.) On the other hand, many of the more recent units are less complex, smaller, cheaper, and designed for much smaller batches. This type of FMS is encroaching on the (former) domain of job shops. There is evidence of a "gap" between the two types, in which the economic benefits of FMSs are less clear (*Figure 5.13*).

## 5.8   CAD

During the early years of CAD, extraordinary returns on investment were sometimes achieved by sophisticated users. For instance, Edward Nilson of Pratt and Whitney Aircraft, a major producer of jet engines, was quoted in *Fortune* (October 6, 1981): "Thanks to CAD [computer aided design] we have gained a 5-to-1 or 6-to-1 reduction in labor and at least 2-to-1 reduction

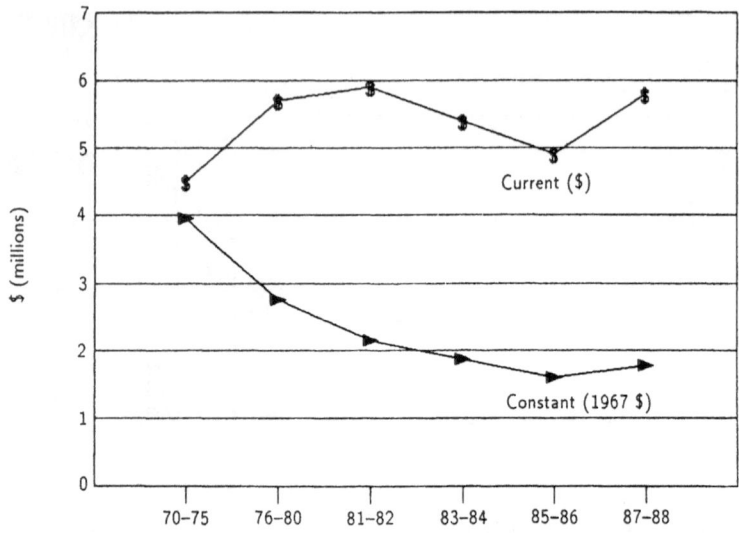

**Figure 5.12.** Average installed cost of FMS. Source: Tchijov, forthcoming.

in lead time. And these ratios go up as high as 30-to-1 and 50-to-1 when we link CAD to CAM."

Based on data gathered by the Institute for Industrial Research, packaged CAD systems, on the average, produced 100% return on investment (ROI) for successful users, and the survival rate was 80% (Harvey, 1983). Productivity gains obtained by substituting a CAD system for a drawing board range from 2 to 3 ×. For the design of printed circuit boards and "chips," the factor of improvement is of the order of 4 to 5 × (Åstebro, forthcoming). Other benefits include reduction in design lead time, ability to design more complex items (e.g., chips), and superior designs. These benefits often result in 15%–20% higher sales.

The CAD industry has grown rapidly, passing the $25 million mark in 1977 and the $350 million level in 1979. (Before 1980 virtually all CAD system producers were in the USA.) Worldwide demand continued to grow rapidly, from $592 million in 1980 to an estimated $2.8 billion in 1982 and

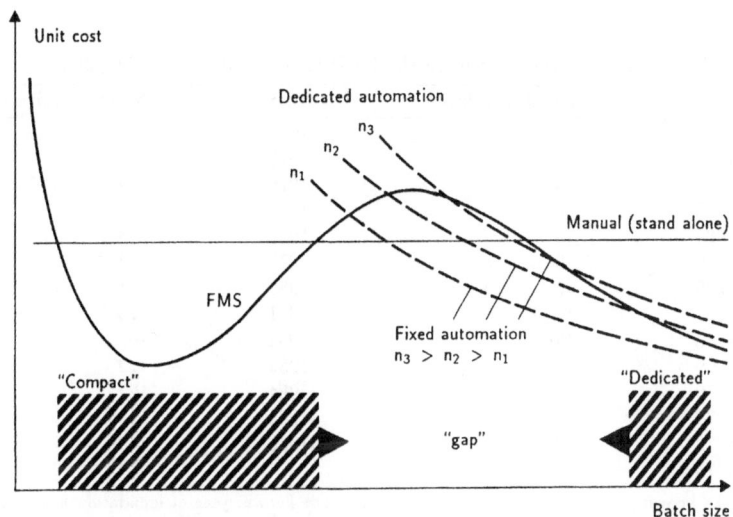

**Figure 5.13.** Two types of FMSs, as function of batch size. Source: Ranta and Tchijov, 1990.

$3.5 billion in 1985 (of which $2.8 billion was supplied by US firms). The number of firms supplying CAD systems is now over 100 *Table 5.8.*

Unit prices of CAD systems are dropping rapidly, as might be expected. The average CAD system installed in 1980 cost close to $500,000 million when 1,500 systems were installed. In 1985, 11,000 were installed at an average cost of just under $400,000. Most of these systems use 32-bit minicomputers. There were about 18,000 CAD installations in the USA in 1985, and probably 25,000 worldwide, with a total of 152,000 work stations. The total number of seats doubled from 1984 to 1985 and again to 1986, the last year for which we have data (*Table 5.9*). The penetration of CAD seats in the manufacturing industry is shown by country in *Table 5.10.*

It is expected that unit prices of systems sold in 1995 will be about 20% of current prices, with 70% of the performance. This is due to the increasing use of CAD adapted for 16-bit personal computers (PCs). It is estimated that 90% of CAD systems will be on 16-bit PCs by 1990 (Ebel and Ulrich, 1987).

**Table 5.8.** Number of firms in the CAD industry during 1960–1986.

| Year | Number of firms | Year | Number of firms |
|------|-----------------|------|-----------------|
| 1960 | 4  | 1974 | 35  |
| 1961 | 5  | 1975 | 38  |
| 1962 | 6  | 1976 | 38  |
| 1963 | 8  | 1977 | 39  |
| 1964 | 9  | 1978 | 45  |
| 1965 | 9  | 1979 | 53  |
| 1966 | 10 | 1980 | 61  |
| 1967 | 14 | 1981 | 74  |
| 1968 | 15 | 1982 | 102 |
| 1969 | 24 | 1983 | 122 |
| 1970 | 26 | 1984 | 122 |
| 1971 | 28 | 1985 | 119 |
| 1972 | 30 | 1986 | 119 |
| 1973 | 33 |      |     |

The known year of entry has been used. When not known, year of foundation has been used. The number of firms is calculated as the accumulated # entries − (# exists + # acquisitions + # mergers). Source: Åstebro, forthcoming.

**Table 5.9.** Market size 1977–1988.

| Year | Market size million US$ | World market size # of installed seats |
|------|-------------------------|----------------------------------------|
| 1977 | 100   | 5,300   |
| 1978 | 200   | 6,800   |
| 1979 | 300   | 8,000   |
| 1980 | 600   | 17,600  |
| 1981 | 700   | 28,700  |
| 1982 | 1,000 | 47,000  |
| 1983 | 1,300 | 60,000  |
| 1984 | 1,800 | 76,500  |
| 1985 | 2,900 | 152,000 |
| 1986 | 3,300 | 303,000 |
| 1987 | 4,100 | n.a.    |

Source: Åstebro, forthcoming.

**Table 5.10.** The penetration of CAD seats in the manufacturing industry.

| Country | CAD per employees (per thousand) | Country | CAD per employees (per thousand) |
|---------|---------|---------|---------|
| USA | 2.87 | Japan | 0.72 |
| Canada | 3.47 | Korea | 0.12 |
| UK | 3.04 | Singapore | 0.99 |
| FRG | 2.46 | Taiwan | 0.52 |
| France | 2.88 | Yugoslavia | 0.10 |
| Italy | 0.31 | Bulgaria | 0.08 |
| Sweden | 3.73 | USSR | 0.02 |
| Norway | 2.58 | India | 0.06 |
| Finland | 1.13 | Argentina | 0.19 |
| Denmark | 1.45 | Brazil | 0.04 |
| Iceland | 0.59 | | |

Source: Yearbook of Labour Statistics, ILO, 1988; Statistical Yearbook of the Republic of China, 1986.

## Notes

[1] This effect is reflected in the long-term rise in "labor share" of output.
[2] Admittedly, this relationship is strongly dependent on dollar–yen exchange rates. For 1984 exchange rates, the USA had an anomalously low rate of robot use and high apparent wage rate, while the reverse was true for Japan. Conversely, assuming 1988 exchange rates, robot use in the USA was essentially average. (See Tani, 1989; and Volume III.)

# Chapter 6

# Organization and Management

## 6.1 Introduction

There is a "chicken–egg" relationship between technological innovations of the kind discussed in Chapter 4 and the organizational changes that (according to one's perspective) either precede or follow. Probably the first important organizational change in the history of manufacturing was the factory system itself. Prior to the nineteenth century the predominant unit of production was the family unit, perhaps supplemented by a few apprentices or hired assistants. Large-scale production was rare. There were very few "factories" in anything like the modern sense. The division of labor into tasks was therefore quite slow to evolve. When Adam Smith popularized the idea as applied to pin making in *Wealth of Nations* (1776), he based his observations largely on time and motion studies carried out by the Frenchman Perronet in 1760. Perronet concluded that the standard rate of output (for No. 6 pins) should be 494 pins per hour of labor. A similar study carried out in the 1820s by the English mathematician Charles Babbage concluded that the appropriate standard (for No. 11 pins) should be 721 pins per hour.[1]

These ideas were finally systematized by the work of Frederick W. Taylor over the 30-year period from 1881 to 1911. Taylor, an engineer working at Midvale Steel Co., classified work into *tasks* and *task elements*, which were individually timed. Taylor argued that this sort of analysis was an important basis for improving work methods. However, at first both workers and management regarded the Taylor system as a device for setting "piece rates," which management feared would be set too conservatively while workers feared the opposite.

Eventually, Taylor's work on "scientific management" found a better reception on the part of management at least. By 1917 hundreds of plants in the USA had introduced variations of the Taylor system. The most influential of them, of course, was Henry Ford. Ford's spectacular success was widely attributed to Taylor's principles.

The Taylor system (or Taylor–Ford system) was based on a fundamental assumption that seemed obvious and therefore received little critical attention. This organizational assumption – implicit in the notion of *division of labor* – was that all of the individual task elements in a job are *independent* and *separable*.

It follows automatically from this notion of independence and separability that organizations can (and should) be structured hierarchically in a "command and control" structure with organizational components corresponding to tasks. It also follows that each organizational unit – down to the machine operator on the factory floor – can and *should* ignore the other units in its work. Independence and separability imply that the output of the organization as a whole will be maximized by maximizing the output of each organizational unit, making use of Taylor's methods of analysis and optimization. As will be emphasized later, the assumption of independence and separability of individual activities in a complex, modern manufacturing operation is faulty. The organizational implications of this recognition are only now beginning to emerge.

In the case of computers, it can hardly be argued that organizational changes made them possible (or played any significant role). Unquestionably computers came first, and organizations have adopted them and adapted to them – more or less. (In fact, I would argue, rather less than more, at least to date.)

In the case of computer integration in manufacturing (CIM), however, the technology, in the broadest sense, is indistinguishable from its use. The key, after all, is *functional integration*. The ultimate goal is a manufacturing system in which machines are "smart" and can "talk" directly to each other, bypassing most human interfaces that characterize the present system. From an organizational perspective, one might ask whether such a utopian scheme needs to be discussed in (human) organizational terms at all. Where do the humans fit into the "factory of the future" – if at all?

In the rather long run, to be sure, there will indeed be a fairly drastic reduction in the number of human workers who are directly involved with production in a hands-on sense. This topic is reserved for Chapter 8. For the foreseeable future, however, humans will continue to be involved in the manufacturing process.

Among the significant organizational innovations that have appeared on the manufacturing scene in the past 40 years, one must mention statistical quality control (SQC), which has evolved into something called total quality control (TQC), group technology (GT), materials resource planning (MRP), and just-in-time (JIT).

## 6.2 SQC and TQC

The first of these "modern" organizational innovations (TQC) began as a factory-level statistical monitoring technique known as statistical quality control (SQC) that was initially pioneered by W.A. Shewhart of Bell Telephone Laboratories in 1931 and applied in the plants of Western Electric Co., manufacturing arm of American Telephone & Telegraph Co. (AT&T). The basic ideas were brought to Japan by W. Edwards Deming and others during the US military occupation after World War II, to assist the Japanese to get their telephone system operating again. These ideas were widely adopted in Japanese industry, and Deming became far better known and honored in Japan than in his native country. In fact, the prestigious Deming Prize for quality control has been awarded annually in Japan since 1951.

The straightforward techniques of statistical quality control (SQC) were extended and elevated to something more akin to a general philosophy of management at all levels – called total quality control (TQC) – by A.V. Feigenbaum of the General Electric Co. (Feigenbaum, 1983). But the TQC approach was adopted far more enthusiastically in Japan than in the USA (Karatsu, 1984; Kume, 1984; Trevor, 1986). In a sense, the TQC philosophy can be regarded as a reaction to Frederick Taylor's single-minded emphasis on subdivision of work into separable tasks, defining each task narrowly and precisely and assigning each task to a worker who need know nothing about any other task. The idea was to maximize the efficiency and hence the output of each worker, individually, without giving due consideration to the potential for interaction and interference (i.e., negative externalities) between tasks. Yet such interactions are inevitable in any system.

In particular, tasks are inescapably interrelated by the fact that the output of one step in a complex sequential manufacturing process becomes the input of the next step. Errors that are not detected and corrected as soon as they occur can become defective parts and defective assemblies. The later in the process the defect is detected, the more expensive it becomes to correct it by "rework." Escalating cost of error detection and error correction in successive stages of fabrication and assembly is a characteristic of increasingly complex products. (The increasing importance of assembly as an element of cost, as a function of growing product complexity, has already been noted.)

A corresponding increase in the importance of inspection and rework – defect control – is behind the growing emphasis on quality improvement in manufacturing. It is easy to see that a Taylorist–Fordist approach to manufacturing, which attempts to maximize throughput at the task level, will tend to permit (indeed, encourage) an excessive number of errors and defects in the product, resulting in unnecessary downstream costs of inspection, repair, and rework. It is because the problem is systemic – rooted in the basic organizational philosophy of Frederick Taylor and Henry Ford – that TQC is really a management philosophy, not a technique.

Nobody can say for certain whether it was Taylor's ideas of task element independence and separability that promoted the extreme specialization of machinery in the mass-production industry, or whether the specialization of machinery was an independent development which merely fit neatly into the Taylor–Ford organizational structure. It does not really matter.

The Japanese, who were less influenced by Taylorism–Fordism than Americans, were quicker to see the implications of quality control as an overall manufacturing philosophy and to adapt their manufacturing organizations to it. In concise terms, Japanese manufacturers emphasize the elimination of defects and "rework." The analogy with Ford's emphasis on the elimination of "fitting" is compelling and appropriate.

The long-run solution is to design a system to exclude the human worker from the operational ("hands-on") part of the production process. Thanks to solid-state monolithic integrated circuits and large-scale integration (LSI, VLSI), modern computers are of the order of 100,000 times less error-prone than human workers (McKenney and McFarlan, 1982). The direction of technological change in manufacturing is slowly, but inexorably, moving toward the substitution of computers and "smart sensors" for human eyes and hands in all of the "on-line" phases of production. But in the short-run, human workers are still needed for many tasks.

In practice, Deming's "14 points" reduce to four main management principles: (1) commitment by upper management; (2) employee education; (3) formation of problem-solving teams across organizational boundaries; and (4) recognition that TQC is a long-term, *continuous process*. As can be seen

**Table 6.1.** Improvements of "R" Corporation, "W" Division, after TQC.

|                       | Before TQC (%) | After TQC (%) |
|-----------------------|----------------|---------------|
| Scrap                 | 15.0           | 7.5           |
| Return/rework         | 1.0            | 0.3           |
| Total cost of quality | 18.9           | 10.9          |

from the vagueness and generality of these principles, the details (and the results) vary enormously from case to case. Often, TQC is introduced in two or three phases:

- *Phase I*: Training of groups of employees in TQC tools and techniques (such as "Quality Circles").
- *Phase II*: Implementation within a plant.
- *Phase III*: Implementation to other plants.

Phase III is *not* an automatic extension of Phase II. The natural temptation in many firms is to try to accelerate the adoption–diffusion process by moving a group of executives from a plant where TQC has been successfully implemented to others. In too many cases the result has been to undermine the continuing commitment of management in the original plant (principle 1).

A number of spectacular successes have been achieved, however. *Table 6.1* lists one example of a plastics parts manufacturer ("R" Corporation, "W" Division) reported improvements. Another instance, a food products firm ("C" Corporation, "FFG" Division) reported the following results: quality index, +23%; consumer complaints, –37%; container inventories, –16%; product recondition/rework, –52%; and product losses, –16%.

Perhaps the most fanatic of the firms newly converted to TQC (or its equivalent) are the US electronics firms that were badly hurt by the Japanese competitive challenge of the early 1980s. Xerox is one such firm. It had been using statistical quality controls since the early 1970s, but it took a major reorganization in 1984 to effect real changes. The company spent $1,300 per employee ($130 million in all) for training in quality control. But it recovered nearly that much ($116 million) in 1988 alone in reduced costs. The defect rate at Xerox has dropped 93%, and market share is up (from the low point in 1984) by nearly 50%.

Hewlett–Packard is another success story. Its companywide "10X" program (to improve quality tenfold) has already saved an estimated $600 million in warranty repairs and $3 billion in manufacturing costs.

The most ambitious TQC program of all is the "Six Sigma" program at Motorola, a serious companywide attempt to achieve near total perfection in manufacturing by 1992.[2] The program was begun in January 1987. It has already saved the company an estimated $250 million per year in manufacturing costs (according to Motorola's new vice president and director of quality), and aims toward a reduction of $800 million per year by 1992.

As will be seen later, the organizational changes characteristic of TQC implementation are very similar to those associated with JIT (and CIM) implementation.

## 6.3   Group Technology (GT)

Whereas TQC and some of the organizational innovations to be discussed later are, in a sense, *reactions* to the influence of F.W. Taylor, GT is a direct outgrowth of Taylor's work. It can be defined as "a manufacturing philosophy or concept that identifies and exploits the sameness or similarity of parts and operation processes in design and manufacturing" (Ham, 1971).

The core of GT is a classification and cooling system for parts "families," i.e., geometrical shapes. Such a system was part of Taylor's System of Management in the early years of the twentieth century. For many years the tendency was for firms to develop their own individual parts classification systems. However, in the 1950s and 1960s some general principles of classification and factory layout began to emerge, and the subject became a subdiscipline of industrial engineering.[3] The combination of formalized coding schemes and computers has made GT an important tool for manufacturing management since the early 1970s.

In its present form, GT can be thought of as a computerized file in which data are stored for all parts (by name and code number), and cross-referenced by shape, tolerance, material, special processes, and subassembly or by products in which the parts are used. A typical coding scheme is illustrated in *Figure 6.1*. Lists of parts with a given characteristic (shape or material) can be compiled to facilitate planning.

One major use of GT is to reduce or eliminate unnecessary duplication of effort in parts design and manufacturing. When a new product is designed, existing part designs can often be utilized. However, in a typical modern metal-working factory, the number of parts designs already on file is in the tens or hundreds of thousands. (For instance, Westinghouse has designs for

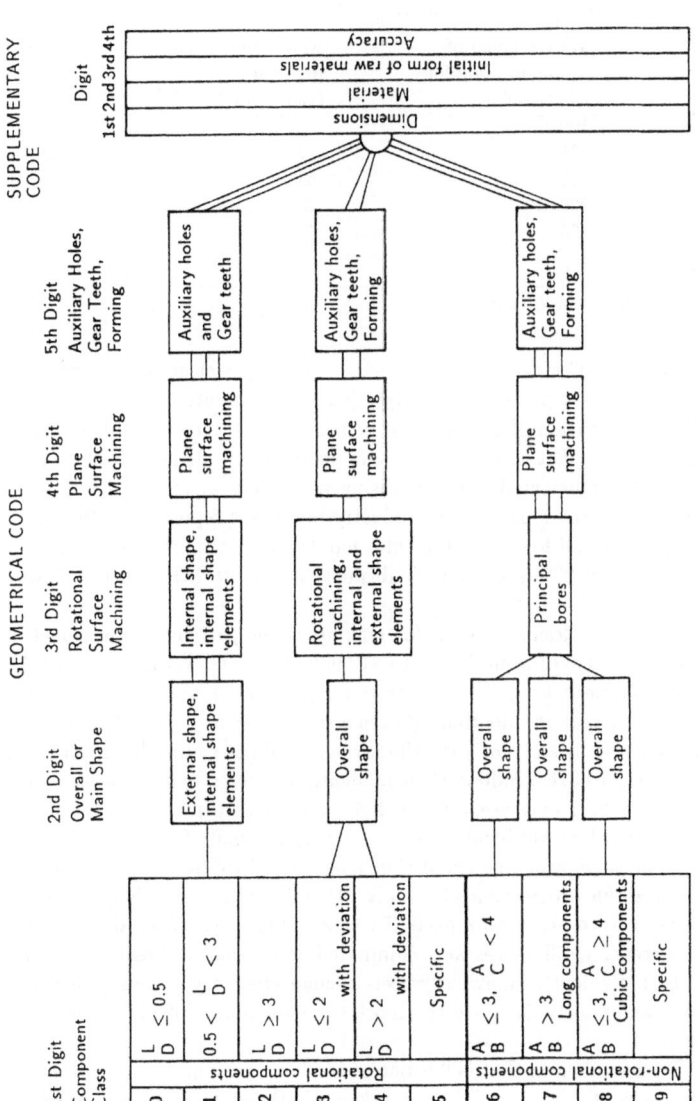

**Figure 6.1.** Group technology workpiece classification. Source: Opitz, 1970.

**Table 6.2.** New parts introduction.

| Firm | Total number of components on file (thousands) | New components per year (thousands) | Growth rate % |
|------|------|------|------|
| A | 100 | 9 | 9 |
| B | 36 | 8 | 22 |
| C | 60 | 29 | 48 |
| D | 220 | 33 | 15 |
| E | 160 | 34 | 21 |
| Total | 576 | 113 | |

Data from: Opitz and Wiendahl, 1970.

more than 50,000 steam turbine wheels.) Finding existing designs "similar enough" for a new purpose is clearly a job for a computer.

The combination of a decreasing lifetime and an increasing diversity of products means that new product – and component – designs must be generated and implemented with increasing frequency. A study of five firms in the West German engineering industry in 1970 is reported in *Table 6.2*. Averaging over all five firms the total number of new parts designs on file was growing by 20% p.a. It is doubtful that this growth rate has slowed down.

The cost for designing tools for making a new part in a typical metalworking shop ranges from $1,300 to $12,000 (Hyer and Wemmerlöv, 1984). The West German firms listed in *Table 6.2* introduced an average of 22,000 new parts per year. Evidentially, if even a small fraction – say 10% – of them can be replaced by an "old" part, the savings would be significant, from $2.85 million (low) to $26.4 million (high). Identifying these opportunities is one purpose of GT. In one case, a single division of an aerospace company found it had designed nearly identical nut and coupling units five different times (ibid). The same firm also found that a list of 2,891 different part numbers in its design files corresponded to only 541 different shapes (ibid). Another firm used GT to analyze its parts list of 20,000 items. When duplications and unnecessary differences were eliminated, fewer than 700 remained (Harvey, 1983). Evidently many early users encountered very serious difficulties. Reliable and timely data are the most common source of implementation problems.

It is fairly obvious that GT is one of the basic prerequisites of computer aided design (CAD). By the same token, CAD has been a great boon to

GT. Indeed, a modern CAD system can automatically classify and code the design. More important, it is possible to eliminate duplication at the design stage.

Another use of GT is in computer aided process planning (CAPP). For each existing part at least one process plan was also generated. With the help of the parts classification systems and codes, manufacturing process planning – once almost entirely the province of experienced machinists – has begun to be formalized also. At the lowest level of sophistication, it is possible to use a GT file to store and retrieve process plans efficiently. Similarities in parts shape suggest natural groupings of parts into families. When the alternative process plans are compared, those machines in common can be selected (also by computer) and unnecessary plans can be eliminated.

For instance, one firm had 477 different process plans developed for 523 different gear shapes. But when the different plans were compared (bearing in mind that each plan can normally be adopted to a number of related parts), it turned out that 400 were redundant. That is, only 77 plans sufficed for all 523 gear shapes (Hyer and Wemmerlöv, 1984). In another case, a firm produced 150 different parts via 87 plans on 51 machines. When the plans were compared, it was found that only 31 plans and 8 machines were sufficient (ibid). Typical benefits of GT reported by users are reduced throughput time, –70%; savings in setup cost, –25%; reduced work in progress, –20%; increased capacity of equipment, +40%; and quality improvements, +75% (Harvey, 1983).

A 1983 survey by the Institute for Industrial Research revealed that firms successfully utilizing GT typically experienced a return on investment (ROI) of 150%, while CAPP users obtained 100%–300% ROI (Harvey, 1983). On the other hand, 20% to 50% of the firms introducing GT did not adopt it on a permanent basis, being satisfied with single-period gains.

## 6.4   Materials Resource Planning (MRP)

At one level of description MRP is a category of packaged computer software for back-scheduling the major events in a production process, given the required date for delivery of the final product to a customer. It is a direct descendent of the so-called *critical path method* (CPM) that was developed into a formal quantitative management tool in the 1950s for the US Department of Defense and (later) NASA.

As applied in manufacturing, MRP is also essentially a philosophy of planning based on *expected* demand. Rightly or wrongly, this is sometimes contrasted with JIT (Section 6.5), which is characterized as a system driven (or pulled) by *real-time* demand. In reality, of course, the two systems overlap in many ways.

The basic purpose of both MRP and JIT is to reduce (or eliminate) unnecessary inventory or work in progress. In the case of MRP, this is done by forecasting future demand and working out its implications for current activities on a daily or weekly basis, taking into account current backlog, sales, inventories, and management priorities. The system generates a master production schedule and automatically prepares shop orders and purchase orders for new materials. In the first generation system (MRP I) this was normally updated on a weekly basis. One of the first firms to adopt the system companywide was Black & Decker (US) around 1970. In its early versions it could be characterized as a "push" system, inasmuch as demand for the coming week was assumed and any deviations from the projections had to be covered by fluctuations in inventory. In the current system (MRP II), it is updated daily.

The basic information flow scheme of MRP is shown in *Figure 6.2*. Effectiveness, in practice, depends on the details of implementation. In the 1970s, MRP systems were typically computer-limited. Today, this is no longer the case. The accuracy and timeliness of data inputs, especially feedback loops, are the most critical factors. Unfortunately, many of the data input nodes are still controlled by human "information transducers," with a tendency to make mistakes. Because of the data reliability problem, in turn, many firms have been reluctant to depend on MRP. (They often ran MRP I in parallel with traditional judgment-based planning systems.) Presently, users tend to rely more upon MRP II with the shorter turnaround times than they did in earlier years.

MRP has apparently been a good investment for some users. The Institute for Industrial Research found that successful adopters were achieving an ROI of 100%. However, only 25% of the systems originally installed were still in use, as of 1983 (Harvey, 1983). Lack of reliable and timely data seems to have been the most common source of difficulties. New hardware (such as scanners) for automatic data entry is fortunately becoming more widespread. This greatly increases both the speed of updating and the reliability of the data inputs.

MRP in some form is now widely utilized in large manufacturing establishments. By 1987 it was also being used by 40% of mid-sized (US)

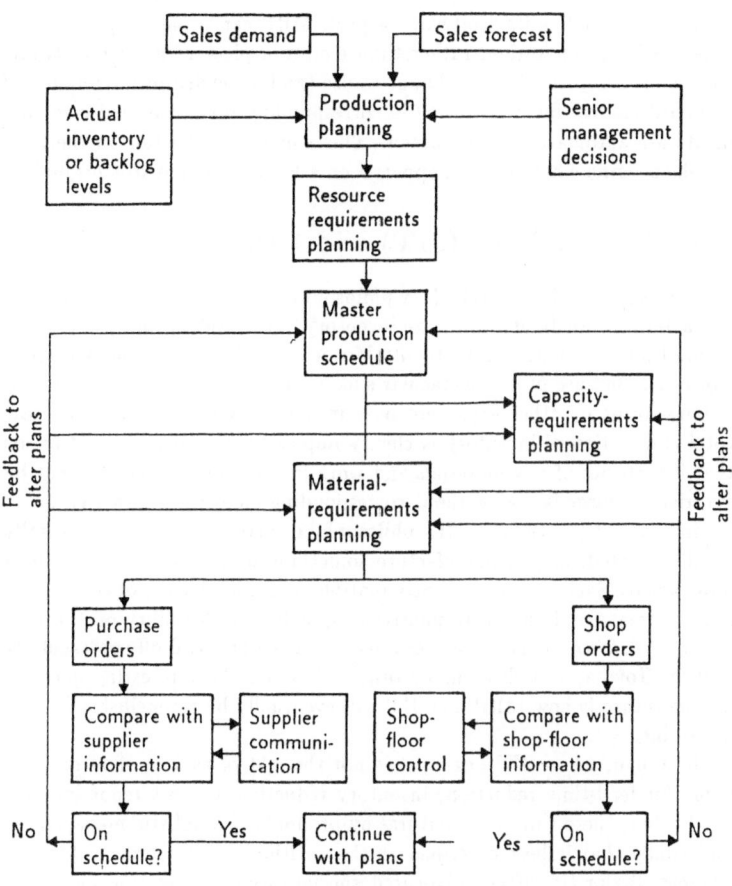

**Figure 6.2.** Flowchart of manufacturing resource planning. Source: Gunn, 1982.

manufacturing sites (100–1,000 workers). Market studies indicated that the worldwide MRP software market was over $800 million in 1987, and expected to reach $1 billion in 1988, with similar growth projected through 1991 (Dataquest, 1987).

Notwithstanding its "software" aspect, MRP can properly be called an organizational innovation, rather than simply a piece of generic software. This is because, once it is really put to use (and depended on), a number of standard clerical functions and the corresponding organizational compartments are automatically eliminated. This wholesale reduction in the need for office workers has not yet happened on a large scale, but it is beginning.

## 6.5   Just-in-Time (JIT) and CIM

"Just-in-time" delivery (JIT) is a philosophy of management that can be characterized briefly as "eliminate inventory." It is an organizational innovation that was pioneered by Toyota Motor Co. (where it became known as *kan ban*). Because of its spectacular success, it has been widely adopted in various forms by other firms, not only in Japan, but elsewhere. Whereas total elimination of inventory is clearly impossible in most cases, it seems that inventories of raw materials and work in progress can often be cut by surprisingly large factors without corresponding increases in other costs.

In contrast to MRP, the JIT philosophy of management is to be totally "pull" oriented, i.e., to manufacture something only when there is a firm order for it. Such a system is incompatible with forecasting, of course, but it puts great emphasis on responsiveness, skill, and flexibility on the part of the work force. This human-resource approach has paid off spectacularly well for Toyota, as well as many other adopters. An interesting question for the future is how MRP and JIT will eventually be reconciled, as seems inevitable.

In Japan, where JIT began, it is not thought of as a method or technique for lead-time reduction, inventory reduction, or cost reduction, *per se*. Rather, these things are natural by-products of a holistic management philosophy that is best described as the antithesis of Taylorism. That is, whereas Taylor (in effect) advocated specialization by function and independence of operation within each function, the JIT philosophy emphasizes the primary goal of product quality and the critical importance of employee and customer involvement across organizational boundaries. "To achieve this, the ideal is for every employee to became a little bit of an engineer, designer, quality expert, and marketeer, while still an expert at his own job" (Hall, 1987). In this respect, the JIT program is similar if not identical to both the TQC program and the CIM program.

However, in practice, the primary focus of JIT is on reducing lead time. Advantages of doing this successfully are many: less waste, fewer inspections, less material movement, less inventory, faster response to customers. But to achieve this invariably involves a comprehensive rethinking of the firm's approach to manufacturing. By contrast, the primary focus of CIM is on flexibility and functional integration.

Bearing in mind that JIT is a *philosophy*, not a *system*, it is nevertheless worth citing examples of its benefits. One interesting case history is the "MM" Division of "B" Corporation, a US manufacturer of outboard engines. In 1982 it was forced to end a joint venture with a Japanese firm and decide between offshore procurement and domestic production. It opted for the latter and made an explicit decision to move away from traditional manufacturing methods and toward a JIT-oriented, flexible manufacturing plant.

The first step was an application of GT: regrouping the machines into short lines (U-lines), based on parts families. Each line is operated by a team. These U-lines are obvious candidates for FMCs or FMSs in the future, but major benefits were achieved even *without* extensive use of computer automation. The benefits of JIT included work in progress reduction, −67%; floor-space reduction, −50%; machining labor reduction, −25%; and quality index, +30%−60%.

Another example from the USA is the well-known Harley−Davidson Motorcycle Co. In 1982 it, too, was forced to undertake a drastic revamping of its manufacturing process as a condition of survival. It did so by adopting the JIT philosophy, with particular emphasis on quality improvement.

On the implementation side, it initiated extensive quality and productivity programs. About half of the employees now belong to voluntary "quality circles." Grievances were halved, and absenteeism dropped 44%. With respect to manufacturing philosophy, Harley−Davidson created its own version of JIT called "materials-as-needed" (MAN). Overall inventory dropped by 50% from 1982 to 1986, a savings of $20 million (as compared with $3 million net earnings for FY 1985). By restructuring the manufacturing process (using GT) and standardizing tools, dies, and fixtures, setup times were cut by 75%. Quality improvements have also been dramatic, enabling the firm to double its warranty period with 100% coverage on both parts and labor.

## 6.6    Meta Systems and Protocols

The next major step in manufacturing technology is integrating CAD/CAM
and FMS technologies. Very briefly stated, the final breakthrough will eliminate the routine human interfaces between the "office" functions, such as
order taking, design, purchasing, inventory control, and scheduling, and the
activities on the shop floor itself. Many of these individual functions have
been computerized. But the individual computer programs were (in most
cases) never designed to "talk to each other" directly.

Until relatively recently, there was really no question of avoiding the
use of humans as information input-output "transducers." Computers only
perform the functions they are programmed for, and programs (until the
present time) are specialized. They only deal with information that is prepared and formatted in special ways. The electronic output of one program
is rarely in a form that can be fed directly into another program without
adaptation and modification – often drastic. Thus, it is typically reduced to
hard copy (i.e., paper) and processed by a clerk or technician. An index of
progress toward the final step in computer integration will be the gradual
elimination of paper records. The electronic repository of the operational
information that permits the business to function is the computerized companywide management information system or MIS.

The MIS is only one component of a larger meta-system comprising sensors, active and passive data files, a communication network to retrieve data
and send it where and when it is needed, and a set of operating programs.
This description could apply, of course, to any computer and (with minor
modifications) to any animal. (The information manager is the brain, the
sensors are the eyes and ears, the files are the memory, and the communications network is the analog of the central nervous system – CNS.)

Each firm must have all of these functions, of course, but most of the
interface nodes are presently still human. In particular, the memory of a
firm is normally divided into a large number of distinct files (many not even
computerized) and often accessible only by file clerks using manual means
of retrieval. Similarly, the CNS of a firm typically consists of a number of
systems operating in parallel including face-to-face conversations, telephone
conversations, internal mail (for hard copy), E-mail, and electronic data
lines.

As the paper files of correspondence, records, and blueprints are increasingly replaced by magnetic tapes or disk storage, the transcription process
(from paper to electronic format) is increasingly critical and is, increasingly,

a bottleneck in the system. Scanners and "smart" sensors are increasingly able to replace humans in this function.

The other bottleneck is the lack of a universal language permitting computers and electronic controllers made by different manufacturers to send and receive intelligible messages to and from each other. The present situation is, essentially, a "tower of Babel" with competing and incompatible operating systems (e.g., MS/DOS, UNIX) programming languages and data formats from different manufacturers.

The long-term solution may be agreement on national and international standards, which would limit the number of options and permit software to be written with enough generality to be truly portable (as packaged PC software is today). The best-known examples, of course, are GM's Manufacturing Applications Protocol (MAP) and the Communications Network for Manufacturing Applications (CNMA), being developed under the EEC's ESPRIT program. In the interim, each firm must essentially create its own internal system (and standards).

The planning and creation of such a system are major steps for any company. To gain some appreciation of its magnitude, the well-documented case of Ingersoll Milling Machine Co. is worth recapitulating. In the late 1970s, Ingersoll reviewed its internal software situation and counted 1,300 application programs, using 225 data files, much of it redundant and uncoordinated. Many programs were not used to process data, but simply to move it. Top management concluded that this mode of operation could not continue. Over a period of two years, all the old programs were replaced by (or embodied in) a single integrated corporate-level management information system (MIS) and data base management system (DBMS). Its modules are as follows:

- Master schedule.
- Engineering design
    - Assembly and piece part drawings
    - Bills of material.
- Production planning and control.
- Inventory control.
- Purchasing and accounts payable.
- Routing and process planning.
- Numerical control programming and post processing.
- Flexible machining system

- – Parts storage and retrieval
- – Automatic transportation
- – Part identification and tracking
- – Direct numerical control (DNC)
- – Computer aided quality assurance control (CAQ)
- – Tool and fixture management
- – Process data management and report.
- Assembly.
- Job cost and management reports.

Only after the MIS and DBMS were in place did Ingersoll embark on the design of its own FMS, one of the most advanced in the industry.

Quite obviously the introduction of a comprehensive companywide MIS and DBMS, with all that it entails, will have major impacts on management. Not all of them can be foreseen, but some reasonable projections are possible. The following is taken from the study conducted by Jaikumar (1986) of one of the world's most successful small arms manufacturers, Beretta, Italy.

> While Beretta has not as yet integrated its CAD/CAM and FMS tech-
> nologies (i.e., has not moved to computer integrated manufacturing, or CIM)
> a few things are already clear. The organization of work will see another
> radical shift. While the engineering ethos in the NC era was one of systems
> science, in the CIM era we see knowledge and the management of intelli-
> gence as the primary domain of activity. The professional work station has
> replaced the simple electronic gauge. The creation of new products and
> processes, being versatile, is the primary driver. The ability to generalize
> and abstract from experience in order to create new products is the skill
> required. As Beretta has thus far built only one FMS cell and designed
> but a single locking mechanism using CAD/CAM, this may seem like idle
> speculation. Yet evidence from Japanese systems operating unattended at
> night suggest that this is clearly the direction. My own studies of more than
> one hundred FMS systems operating worldwide corroborate this evidence.
> The following are among the conclusions drawn from that research.
>
> The management of FMS technology is taking place in a different man-
> ufacturing environment, and thus consists of new imperatives.
>
> *Build small, cohesive teams.* Very small groups of highly skilled gener-
> alists show a remarkable propensity to succeed.
>
> *Manage process improvement, not just output.* FMS technology funda-
> mentally alters the economics of production by drastically reducing variable
> labor costs. When these costs are low, little can be gained by reducing them
> further. The challenge is to develop and manage physical and intellectual
> assets, not the production of goods. Choosing projects that develop in-
> tellectual and physical assets is more important than monitoring the costs

of day-to-day operations. Old-fashioned, sweat-of-the-brow manufacturing effort is now less important than system design and team organization.

*Broaden the role of engineering management to include manufacturing.* The use of small, technologically proficient teams to design, run, and improve FMS operations signals a shift in focus from managing people to managing knowledge, from controlling variable costs to managing fixed costs, and from production planning to project selection. This shift gives engineering the line responsibilities that have long been the province of manufacturing.

*Treat manufacturing as a service.* In an untended FMS environment, all of the tools and software programs required to make a part have to be created before the first unit is produced. While the same is true of typical parts and assembly operations, the difference in an FMS is that there are no allowances for in-the-line, people-intensive adjustments. As a result, competitive success increasingly depends on management's ability to anticipate and respond quickly to changing market needs. With FMS technology, even a small, specialized operation can accommodate shifts in demand. Manufacturing now responds much like a professional service industry, customizing its offerings to the preferences of special market segments.

We have come full circle. The new manufacturing environment looks remarkably similar to the world of Maudslay; only the world of work has changed. The holy grail of a manufacturing science begun in the early 1800s and carried on with religious fervor by Taylor in the early 1900s is, with the dawning of the twenty-first century, finally within grasp.

The reader is referred to Ranta *et al.* (forthcoming, Volume V) where the Beretta Story is told in full.

## 6.7 New Approaches to Cost Accounting

One of the reasons for the pervasiveness of the management and organization approach that has been called "Taylorism" is that it has become embedded in practices and procedures that (at first sight) bear no direct relation to mass-production technology, or the division of labor. Accounting practices, as they have evolved since the heyday of "scientific management" (1880–1915), still reflect long obsolete assumptions about the production process. The most important is that "cost" is identified with blue-collar wages.[4]

Conventional accounting systems were developed at a time when data collection was expensive and unreliable. It was therefore necessary to make the best use of administrative data that was collected for other purposes, such as direct labor time (from time cards submitted to the payroll department), units produced, and material usage. The standard assumption of the Taylor

**Table 6.3.** Summary of IC test performance from 1986 to 1988: values for inventory and scrap, in thousands.

| Year | Defects (ppm) | Throughput (days) | Schedule performance (%) | Inventory (units) | Scrap ($) |
|------|-----------|-----------|-----------|-----------|-----------|
| 1986 | 1,000 | 35 | 85 | 2,000 | 600 |
| 1987 | 500 | 9 | 89 | 288 | 171 |
| 1988 | 270 | 3 | 99 | 120 | 74 |

**Table 6.4.** Financial summary of IC test department from 1986 to 1988.

| Year | Cost per IC tested ($) | Cost per labor-hour($) |
|------|--------------------|--------------------|
| 1986 | 0.50 | 50.00 |
| 1987 | 0.52 | 54.86 |
| 1988 | 0.55 | 101.64 |

system is that each activity in each department is effectively independent of the other and can therefore be evaluated in terms of some "output" per unit of labor and capital input. This ignores the external benefits – and costs – that one activity may generate for others, and has led to more and more misleading reports. (The difficulty of accounting appropriately for R&D is a classic problem. How should the output be measured?) In particular, conventional methods generally fail to document the very real improvements in performance that CIM systems can bring about.

This problem is illustrated by the following case: *Table 6.3* shows the improvements in performance that were achieved from 1986 through 1988 at the integrated circuit (IC) testing facility of a large US-based computer firm (Kaplan, 1989). These results are nothing short of spectacular, although the firm would be the first to admit that there is still plenty of room for further progress. Yet the financial accounting summary prepared by the firm's accounting department told a different story (ibid). It showed a trend of rising unit costs for testing (*Table 6.4*). In fact, this trend had so alarmed the firm's management, that it had opened a new test facility in Singapore to take advantage of lower costs. Only a special study by outsiders convinced the management that the economics of the US location were actually more favorable, and led to a reversal of the decision (ibid).

The problem with old-fashioned accounting stems from two sources. First, as noted already, it is assumed (usually without question) that the performance of each functional unit in the system can be measured by the same criteria. In the case above, it was assumed that "costs per unit inspected" was a reasonable measure of performance for the facility. The fact that better testing could lead to better manufacturing methods that would reduce the *need* for testing cannot be deduced from the accounting methodology, although any sensible person might realize it. But, of course, "management by the numbers" tends to override common sense.[5]

Extrapolating from this case to general practice, standard accounting methodology fails to admit the possibility that the firm might be better off *as a whole* if some of its interactive support activities were expanded, at the expense of that activity *per se* looking worse in the accounting sense. R&D is one obvious example, yet many firms have given in to their accountants by trying to force R&D to pay for itself and become a "profit center." It is clear that TQC, GT, MRP, and JIT are other activities that pay for themselves only by making the firm as a whole function better. This is also true of accounting. However accountants seldom measure their own performance according to the same yardsticks they routinely use on others.

The second fundamental problem with traditional accounting methodology is its allocational rigidity. The methodology has been standardized for a long time, and it is taught largely by rote. Standard classifications are defined and activities are forced into narrow taxonomic boxes. Everything in the box is then treated as if it were an average of the whole box. In particular, certain activities – such as general management, finance, accounting, personnel, public relations, marketing, and R&D – are treated as "overhead." There may be several layers of overhead, e.g., for a division or a line or the firm as a whole. The overhead is then allocated to each item within the box (division, line, etc.) in proportion to its *direct* (i.e., *variable*) cost.

The problem is that internal differences within the accounting unit (box) are ignored and washed out. For example, the production cost for an item that is made correctly the first time is treated as the same as the production cost for a defective unit that had to be "reworked." This is perfectly reasonable as a basis for pricing, but it is quite misleading as a basis for allocating resources within the manufacturing operation. To take another example, the traditional accounting methodology as applied at the plant level misallocates internal overhead costs – selling costs, setup time, and so on – in proportion to variable costs (hence volume). Kaplan notes:

> Product variants whose selling price exceeded short-run variable costs (again typically measured only by direct materials and direct labor) were considered desirable because they could absorb overhead costs and contribute to

profits. The cost, however, of the added support resources needed to handle the proliferation of products, models and options was not traced to these additional items. Rather, the costs... were allocated across all products, based on their relative volume of production, not on the demands individual products made on the plants indirect resources (Kaplan, 1989, p. 245).

The consequences of these misleading signals were – and still are – that companies often fail to see, and hence to take advantage of, the full benefits of flexibility and CIM. A new approach, known as the activity-based cost (ABC) system, may eventually find more acceptance in industry. The basic idea is to estimate manufacturing costs as a function of volume (batch size), precision, complexity, and the other variables that would appear in a (fully elaborated) cost model such as the one sketched briefly in Section 3.2.

With the help of such a model in operation in the accounting system, true costs are more accurately determined and internal resources can be more appropriately allocated. As an illustration, the ABC system was tried on a Siemens electric motor plant. Instead of allocating all indirect costs over all motors produced in proportion to variable costs, indirect costs were allocated as actually incurred. The result was an *increase* in total costs of 30%–40% for customized motors produced in lot sizes of one, and a *decrease* of 7%–9% in costs assigned to standard motors produced in lots of 100 or more. (This can be a significant competitive difference in a relatively mature price-sensitive market.)

An international consortium of automation equipment manufacturers, multinationals, and accountants, known as Computer Aided Manufacturing-International (CAM-I) advocates a more radical approach. Supporters like Peter Drucker argue that the traditional direct labor-based system cannot be reformed (Drucker, 1990). Instead (it is argued) the standard system should be replaced by a new system in which the basic unit of measurement is elapsed time. There would no longer be a distinction between "fixed" and "variable" costs. Inventory would no longer be treated as an "asset" but as a "sunk cost." "Benefit" is whatever reduces the time taken by a process. Whether such a radical change would introduce new problems – faster need not be better or cheaper – remains to be seen. Stay tuned!

## 6.8   Flexible Management

There is a large management literature on this subject to which I can scarcely hope to add. Certainly almost all of my observations on this subject will repeat what has already been said many times. Nevertheless, it may be worthwhile to summarize some of the lessons for managing a flexible manufacturing company in the age of CIM.

To make the contrasts and the lessons as clear and pointed as possible, let us suppose that "the firm" is an old, established one in the engineering sector organized on the Taylorist–Fordist model, that the former management team has retired (voluntarily or otherwise) to play golf, and that the firm is experiencing competitive difficulties. Assume, too, that the board of directors wants to keep the firm in business, not to liquidate it. I list and comment on some obvious topics of interest to the board of directors, the chief executive officer, and the new management team.

*Qualifications*: What kind of background should the new CEO have? He or she need not be an expert in manufacturing or in the product technology, but the CEO must be a well-educated generalist – with enough knowledge to read and understand every chapter of this book, for instance. The primary function of a CEO of a firm that plans to stay in business is to articulate the *goals* of the company in a way that makes it clear to every employee how his or her contribution is needed. From one point of view the problem is communication. But an actor's rhetorical skills are not enough. (If need be, one can hire an actor.) It is easy to articulate a broad goal, but quite difficult to understand and explain to others exactly how every component of a complex firm contributes to meeting that goal.

To achieve the latter, a considerable technical knowledge is needed. Since the CEO cannot be expected to have all of the necessary technical knowledge, he or she needs one skill above all: the ability to find and recognize people who do have the necessary specialized knowledge, to listen to them, to hear and understand them, and to "translate" the essentials into language accessible to others.

In recent years financial expertise has become arguably the most popular path to the chief executive's seat. To the extent that this expertise has been acquired in business schools or from textbooks, it may well be counterproductive. Certainly, the notion that one can manage any large enterprise "by the numbers" – meaning, the numbers generated by the accounting department – is pernicious.

*Goals*: What kinds of goal statements make the most sense for an engineering firm? Is there anything wrong with setting a goal in terms of profitability or growth? Is there any special magic about setting goals in terms of improved quality (TQC), reduced inventory (JIT), or faster turnaround time?

Goal setting has acquired a sort of mystique, which is probably over-emphasized by some management "gurus." The importance of setting a common goal lies in getting everybody in the group – whether it is an office, a sub-department, or a whole firm – to work together. It is a tool for breaking down barriers. In principle, most of the innovative goal formulations that have been tried out in recent years are ultimately aimed at increasing profitability. Hence, it should not matter how the goal is formulated, as long as everybody understands it and understands how they are expected to contribute.

In practice, however, it is clear that some goal formulations are much more appealing – hence motivating – to employees than others. In particular, if a company can identify an area where it is doing poorly, and which is easily quantifiable on a current basis, there is an opportunity for people to see the results of their common efforts reflected very quickly and unambiguously. Quality control or inventory have this characteristic. By contrast, although profitability is quantifiable over time, for a complex modern corporation it is measurable only by an arcane methodology understood in detail only by the tax lawyers and the accounting department. Worse, there is no methodology that anybody could justify against serious objection to determine who has contributed how much to profits.

Finally, given the recent history of extraordinary emphasis on short-term profitability by Wall Street, and the leveraged buyouts and "restructurings" that have resulted, a great many employees at the lower levels of the hierarchy (in US-based firms, at least) are now cynical about profits. Goals set in financial terms are likely to be regarded by most employees as, at best, not relevant to them and, at worst, threatening.

*Incentive structures*: It is a key function of the top management of any firm to create an incentive system within the firm that is truly consistent with meeting the goals. This is one of the most important and least understood issues. Goal setting is not a new idea. The last CEO (and the one before him or her) probably began his or her tenure with some ambitious goals. But for some reason, they were not fulfilled. Of course, the former leader might have failed to communicate his goals properly – a deficiency in rhetorical skills – but more than likely the difficulty lies elsewhere. The grand goals for the firm as a whole, as stated by the former CEO, were not fulfilled (in all probability) because they were not consistent with the narrower, everyday goals of stewards, supervisors, department heads, division heads, and so on.

The new set of goals probably will not be fulfilled, either, unless the employees get the same consistent message, day in and day out, from every department or office with which they interact. This includes (especially) the accountants and financial analysts. The essence of the problem of translating goals into action is *monitoring performance* and creating appropriate incentives – financial or other. The accounting department is primarily responsible for the measurement of performance. But accountants are extremely conservative. They use standardized methods developed in the distant past and enshrined in textbooks and "professional standards."

The new CEO of an old-line manufacturing firm will therefore have to begin by motivating and reeducating the accountants, to induce them to measure the right aspects of performance. Traditional accounting training is probably not sufficient any longer. Industrial engineers and manufacturing engineers (among others) will have to be involved in the process of developing new and better measures. When the accounting department is providing the right *kinds* of information, everybody else will be getting the right signals. If the accountants do not move with the times, nobody else will be able to do their jobs properly.

*Barrier breaking*: The biggest single problem of the old-line Taylorist–Fordist organization is, precisely, that it was created on the premise that labor is infinitely divisible: each activity of the firm can be broken down into separable, *independent* components. The problems of such organizations – excessive rigidity among them – arise from flaws in the premise. Activities that are really interdependent can be separated only at a cost, and, in the modern world, the cost of maintaining artificial barriers between activities is becoming intolerable.

But what is to be done? There is no single optimum solution. Many firms have introduced the practice of creating many *ad hoc* committees or "task forces" with members from different departments. (The creation of new measures of performance by and for the accounting department would be an example of a possible task for such an interdisciplinary, cross-organizational group.) The Scandinavian "team" method of manufacturing – which replaces the assembly line by a team of multi-skilled workers with rotating responsibilities – is a shop-floor version of the task force idea, except that the team is intended to be quasi-permanent. "Matrix management" is a more formal version of the team approach as applied to nonproduction functions. It is a structuring scheme that defines each job in terms of responsibilities in two dimensions (such as permanent specialty and current task or mission).

Unfortunately, one of the original tools of the industrial engineer, the formal "job description," is now a major horizontal barrier to information flow and efficient allocation of effort. It has become codified into many union labor contracts and (in some cases) into laws. Where this has happened, mainly in the USA and both Western and Eastern Europe (usually to "protect" jobs) flexibility and efficiency are minimized. The classic case is the automated line of machines that must be brought to a halt when a fuse blows, because only a union-certified "electrician" is allowed – by union contract – to change the fuse. As a result, there are often hundreds of different job classifications in large plants where only a few would suffice. By contrast, Japanese firms – whose workers are only represented by company unions – are largely free of such restrictions. Elimination of unnecessary job classifications and rules forbidding workers to perform any task not included in their particular job description must be a major objective of management.

Every vertical level of management is also a barrier to information flow. The necessity for a certain number of levels is not in doubt, but it is interesting that some relatively efficient organizations are able to get by with relatively few distinct management levels. For instance, the Catholic Church manages with just five levels, from parish priest to Pope; even allowing for two or three more levels within the individual church brings the total to no more than eight. Universities typically need no more than six or seven levels, down to the janitor.

Why, then, do some large corporations distinguish 20 or 30 levels of management? The explanation seems to be that creating new (and increasingly meaningless) titles sometimes seems to be a cheaper way of rewarding people than paying them. Certainly, the term "vice president," once redolent of power and prestige, now means almost nothing, at least in some sectors. Nonetheless, multiplication of levels of management is extremely bad for effective communication within a firm, because it makes bottom-up communication virtually impossible.

One of the most effective acts a new management team can do in an old-line firm is to eliminate one-half to two-thirds of the unnecessary management levels (and meaningless titles). Another, and more positive, method of improving communication across management levels is to utilize some form of electronic message network in which anybody with an interest in some topic (say, devising better measures of quality) can access others in the firm – regardless of rank or location – with similar interests.

A final barrier, which is perhaps more important in Europe (but varies from country to country), is the social barrier between white-collar and blue-collar worker.[6] There have been deliberate efforts to break this barrier in Scandinavia and West Germany, and of course in the CMEA countries (where a pattern of reverse discrimination was widely practiced, at least until recently). A social barrier can be just as effective at impeding information flow as any other sort, and an alert management will work to eliminate such barriers if they exist.

*Training and Human Resources*: A function that management will increasingly find inescapable is employee training and retraining. The fact that implementation of CIM requires multi-skilled workers has been amply documented (see Ranta *et al.*, forthcoming, Volume V). The demographic fact that labor force entry will be static or declining for the next decade or two is well known. The fact that automation will continue to eliminate certain kinds of jobs, especially those involving the lower-level cognitive skills, is also fairly clear. So, we face a period of increasing labor shortage, combined with increased turnover.

Finally, it is a fact that many of the workers who are entering the work force of the industrialized world are unskilled and even illiterate. In the USA many of these are illegal immigrants from the Caribbean or Central America, or products of the urban ghettos. In Western Europe, they are immigrants from the former British, French, or Dutch colonies, or from Yugoslavia, Turkey, and the Middle East. A flood of new migrants from East Germany, Poland, and even the USSR is likely to flow West during the next decade.

For all of these reasons, adult education and training will be a growing priority. For other reasons, much of the responsibility (though not necessarily the cost) for basic language skills and literacy training will necessarily be taken over by the private sector, i.e., by industry. In Western Europe, the cost of retraining redundant workers is largely borne by government (in the form of tax credits and direct subsidies), and this pattern seems likely to spread. In the meantime, some large US firms are finding it worthwhile to undertake educational initiatives on their own, even without government assistance. For smaller enterprises, the main difficulty is that individuals who have gained new skills and market value at company A's expense are free to take them to company B. Nevertheless, there is plenty of scope for management ingenuity in devising clever and palatable schemes for upgrading the skills of valuable employees and keeping them on the job. Human resources will be an increasing concern of management.

## Notes

[1] Babbage was also the first to invent a workable mechanical computer (the "difference engine") although no operating version was completed during his lifetime – due to manufacturing problems (Shurkin, 1984).

[2] The term sigma here refers to the so-called "standard deviation" of a normal (Gaussian) probability distribution. Assuming parts differ probabilistically from the ideal specification, the probability of a part being within a specified departure from the ideal can be expressed in terms of multiples of sigma. If the probability is "1 sigma" it is 68.26%; "2 sigma" is 95.46%; "3 sigma" is 99.73%, and so on. Six sigma corresponds to 99.9999998% probability or a defect rate of .0000002%.

[3] See, for instance, Mitrafanov, (1966); Opitz, (1970).

[4] In the 1920s blue-collar wages ("direct labor costs") typically accounted for 80% of total costs other than purchased materials. Today, the average for large-scale producers is less than 20% (*Table 6.3*), and the trend is still down toward a norm of 8%–12% (Drucker, 1990).

[5] Kaplan notes the irony that, while accounting systems were producing more and more irrelevant numbers, top management of large firms in the 1960s and 1970s was relying more and more on management "by the numbers." This trend was glorified in the business schools as "portfolio management" and exemplified by the approach taken by some of the conglomerators of the 1970s, notably Henry Singleton of Teledyne.

[6] In countries with a strong military tradition, this obstacle corresponds quite closely to the traditional barrier between the "officer class" (upper) and the "enlisted men" (lower). It is often enforced by other symbols of social superiority-inferiority such as speech patterns (accents) and the like. In the USA the upper class–lower class barrier is much lower for various reasons, including the lack of a military tradition. However, more or less subtle discrimination against racial or ethnic minorities, or against women, is still widespread.

# Chapter 7

# Economic Impacts of CIM

## 7.1 Introduction

Technological change in manufacturing generates economic consequences through several distinct linkages. The five major categories of benefits to the firm were discussed in Chapter 5, in the context of inducements to the adoption of new technology. They are, again: labor saving; capacity augmenting/capital saving; capital sharing/saving; product quality improvement (i.e., defect reduction); and acceleration of product performance improvement. The first three categories tend to reduce unit costs/prices for the product. This, in turn, stimulates demand. The last two items affect demand directly.

Labor saving and capital saving have an immediate and obvious impact on costs, as discussed in Section 3.2 (*Table 3.1*). Labor-saving opportunities are particularly important in small-scale production because direct labor constitutes the largest single cost element (43.7%). Conversely, capital saving is comparatively more important at larger scales of production.

Capacity augmentation has two different effects, depending on circumstances. In a rapidly growing market, it is equivalent to saving capital; it postpones the need to invest in new capacity. In a slow-growing or static environment, on the other hand, capacity augmentation is likely to destabilize the balance between supply and demand in favor of supply. Excess capacity often leads to competitive price cutting, reduced margins, and financial distress. The market is only re-stabilized if demand responds to the lower prices or if somebody drops out, thus rectifying the imbalance.

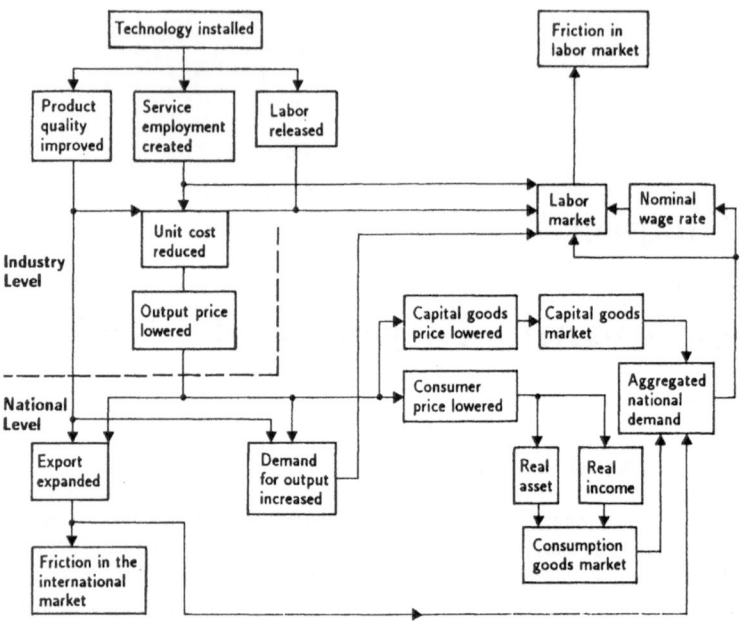

**Figure 7.1.** Economic impacts of technical change in manufacturing. Source: Kaya, 1986.

A general schematic diagram that displays some of the main linkages is shown in *Figure 7.1*. Implications of labor-saving are discussed next, followed by other topics.

## 7.2   Labor and Capital

Lower costs, whether of labor or capital, usually impact on prices which, in turn, drive demand. The amount of the increase in demand for manufactured goods depends on the price elasticity of demand. For most goods, the price elasticity seems to be somewhat greater than unity (in absolute value), whence overall demand can be expected to rise by a modest amount. Price elasticity of demand, $e$, is defined as:

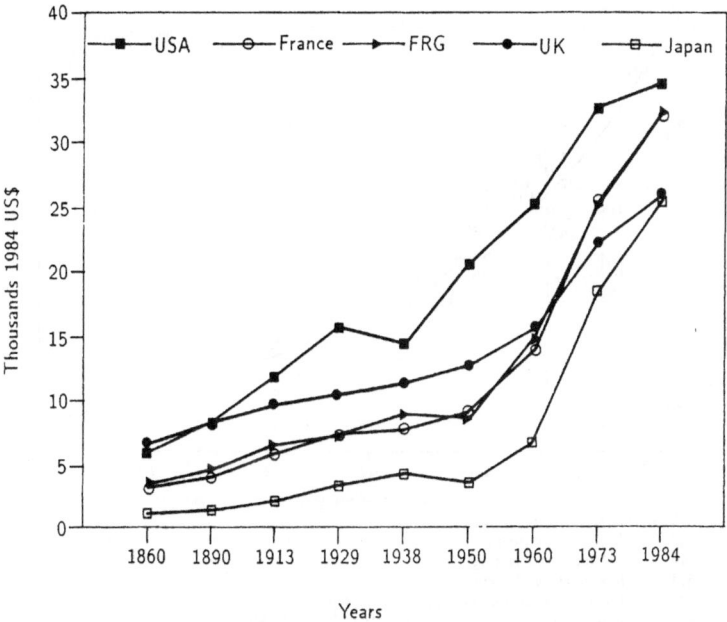

**Figure 7.2.** Labor productivity level, whole economy (GDP per worker). Source: Maddison, 1987.

$$e = -\delta \ln Q / \delta \ln P = -(P/Q)(\delta Q / \delta P) \; , \qquad (7.1)$$

where $Q$ refers to overall demand for manufactured goods and $P$ to the price level (in real terms). Thus, if $e$ has an absolute value of 1.5, it means that a 1% *decrease* in price results in a 1.5% *increase* in sales, corresponding to a 0.5% increase in total revenues.

As noted above, the relative importance of labor and capital costs depends on scale. Historically, most gains in total factor productivity have arisen from labor savings, rather than capital savings. This translates into increased labor productivity, with constant or declining capital productivity, as shown in *Figure 2.16* and *Figure 7.2* (see also Chapter 2). Direct labor is still the largest single element of manufacturing costs in the metal-working

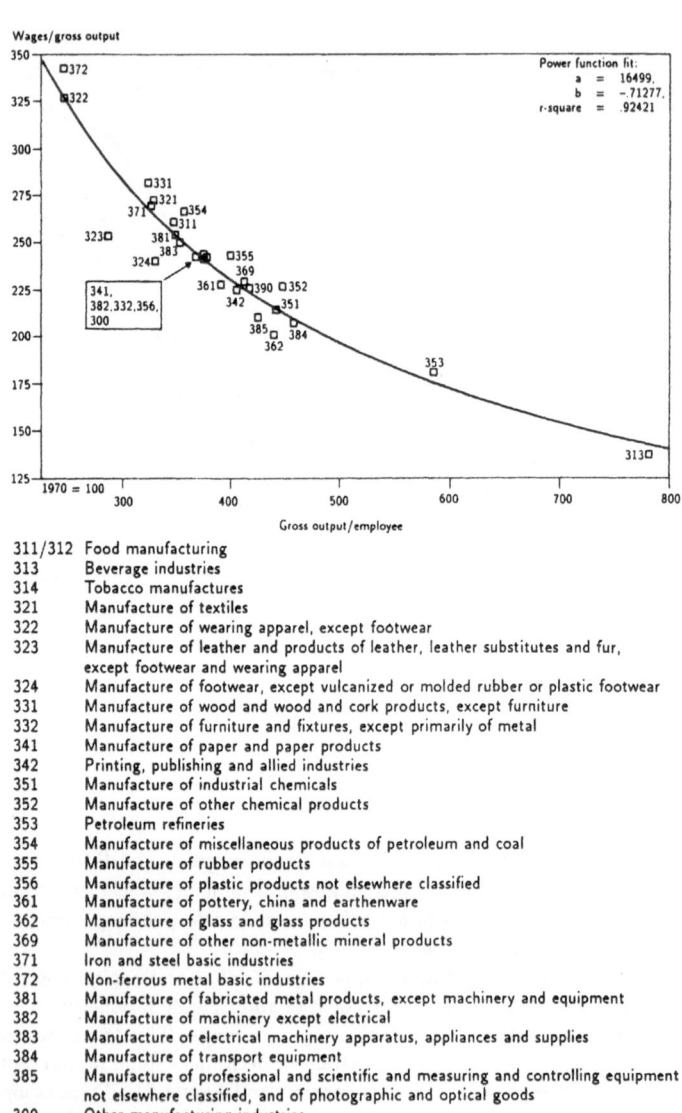

Wages/gross output

Power function fit:
a = 16499,
b = -.71277,
r-square = .92421

1970 = 100

Gross output/employee

| 311/312 | Food manufacturing |
|---|---|
| 313 | Beverage industries |
| 314 | Tobacco manufactures |
| 321 | Manufacture of textiles |
| 322 | Manufacture of wearing apparel, except footwear |
| 323 | Manufacture of leather and products of leather, leather substitutes and fur, except footwear and wearing apparel |
| 324 | Manufacture of footwear, except vulcanized or molded rubber or plastic footwear |
| 331 | Manufacture of wood and wood and cork products, except furniture |
| 332 | Manufacture of furniture and fixtures, except primarily of metal |
| 341 | Manufacture of paper and paper products |
| 342 | Printing, publishing and allied industries |
| 351 | Manufacture of industrial chemicals |
| 352 | Manufacture of other chemical products |
| 353 | Petroleum refineries |
| 354 | Manufacture of miscellaneous products of petroleum and coal |
| 355 | Manufacture of rubber products |
| 356 | Manufacture of plastic products not elsewhere classified |
| 361 | Manufacture of pottery, china and earthenware |
| 362 | Manufacture of glass and glass products |
| 369 | Manufacture of other non-metallic mineral products |
| 371 | Iron and steel basic industries |
| 372 | Non-ferrous metal basic industries |
| 381 | Manufacture of fabricated metal products, except machinery and equipment |
| 382 | Manufacture of machinery except electrical |
| 383 | Manufacture of electrical machinery apparatus, appliances and supplies |
| 384 | Manufacture of transport equipment |
| 385 | Manufacture of professional and scientific and measuring and controlling equipment not elsewhere classified, and of photographic and optical goods |
| 390 | Other manufacturing industries |

**Figure 7.3.** Labor productivity and labor cost, Japan 1986.

sectors, so the reduction of direct labor costs remains a significant target of opportunity.

The impact of increased labor productivity on product price (cost) can be estimated econometrically by taking advantage of the above relationship between labor cost $(C_L)$ as a fraction of output $(C_L/Y)$ and labor productivity or output per worker $(Y/L)$. The power law relationship between these two indices is shown in *Figure 7.3*.

From the graph (based on Japanese data) it can be estimated that an *increase* in labor productivity of 1% will *reduce* the labor cost fraction by about 1.4%. This cost reduction, in turn, permits a corresponding price reduction, roughly in proportion to the labor cost fraction. (This depends on scale of production, as shown in *Tables 3.1* and *3.2*.) If the direct plus overhead labor fraction of total costs is 50%, then a 1.4% decrease in the labor fraction would correspond to an overall cost decrease of 0.7%. Typically, the price elasticity $(e)$ of sectoral output (demand) lies between $+0.3$ and $+0.7$ in most sectors. That is, a price decrease of 1% stimulates a demand increase between 0.3% and 0.7%. In summary, a rise of 1% in labor productivity is likely to yield a cost decrease of 0.5% to 0.6%. This, in turn, will generate added demand in the range of 0.2% to 0.5%.

Needless to say, capital savings would be reflected in changes in demand in exactly the same way. (The consumer is assumed to be indifferent to the source of the saving.)

## 7.3 Scale Dynamics

Economies of scale do not fit well into neoclassical economic theory, which prefers to assume constant returns to scale for most purposes. However, economies of scale do exist (as discussed in some detail in Chapter 3), and they play a major role in driving economic growth. As markets grow in size, leading to increases in the scale of manufacturing, costs tend to decline. In a competitive market this would normally be followed by price cuts. Declining prices, in turn, induce increased demand. (The measure of this effect is called *price elasticity of demand*: when markets are unsaturated this measure tends to be greater than unity, in absolute value, and conversely.)

The sequence of events described above is effectively a chain reaction. It tends to continue until the market for the manufactured product is saturated and the price elasticity of demand falls below unity. The fact that declining prices tend to stimulate increased demand with increased revenues means

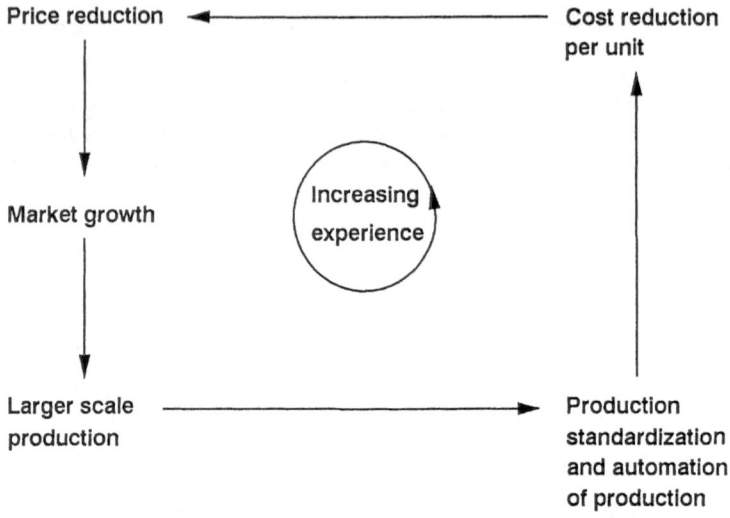

**Figure 7.4.** Cost-price-driven product. Effective in later (adolescent, maturity) stages of the life cycle, subject to price elasticity and returns to scale in production. The latter, in turn, requires standardization, i.e., mass production.

that economies of scale can actually be a major engine of economic growth (Salter, 1960). The chain-reaction mechanism involved is sketched in *Figure 7.4*.

The power of the Salter mechanism depends on the actual numerical value of the average price elasticity, $e$, which we can only estimate very crudely from indirect evidence. Taking all manufactured products together, it seems likely that the price elasticity of demand for most products in the OECD countries is now significantly less than 1.5 and possibly not more than 1, signifying that most markets for goods (notably automobiles and appliances) are approaching saturation. On the other hand, in the developing countries one would expect generally higher price elasticities (i.e., $e$ in the range 2–4) corresponding to less saturated markets.

The chain reaction resulting in increased output and lower prices, based on economies of scale, has contributed significantly to economic growth in the advanced economies during the last 40 years. To account for economic

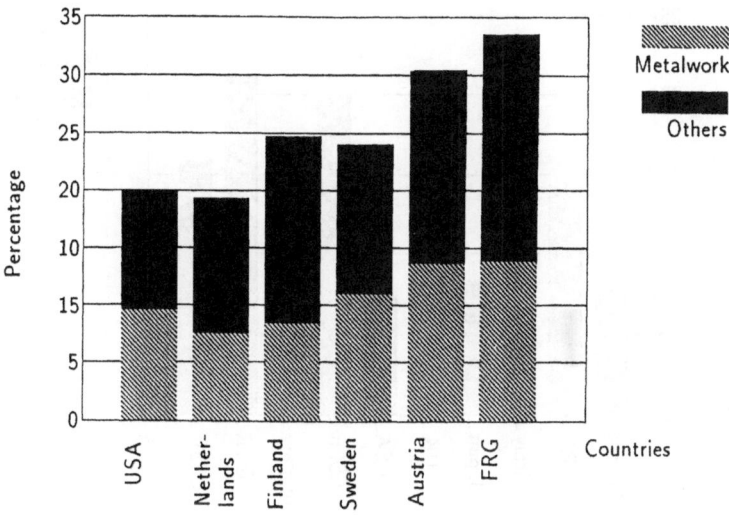

**Figure 7.5.** Manufacturing employees as percentage of total work force.

growth, Edward Denison of the Brookings Institution has carried out a series of empirical studies in which he quantifies the contributions from a variety of exogenous factors (Denison, 1962, 1967, 1974, 1979). Those factors include shifts of rural population into cities, changes in land use, changes in laws and regulations, changes in educational level, capital investment in structures and equipment, changes in inventories, changes in working hours, changes in the age–sex composition of the work force, increases in the size of markets (scale), and increases in "knowledge." (Not all of these factors are positive during any given period; for example, during the 1948–1973 period, working hours declined, contributing negatively to growth. After 1973 almost all factors contributed negatively.)

Without trying to summarize Denison's methodology or results, suffice it to say that during the 1948–1973 period US growth averaged 2.6% per annum (p.a.) in real terms (adjusted for the business cycle), of which Denison attributes 0.4% p.a. to increasing scale. During the 1973–1976 period (when growth was negative at −0.6% p.a.), Denison still attributes a positive contribution of 0.2% p.a. to increasing scale. In short, over the entire period

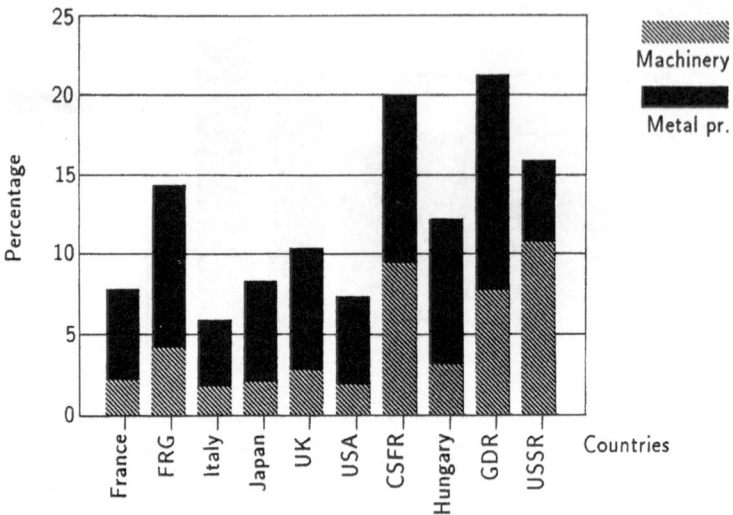

**Figure 7.6.** Machinery manufacturing and metal-products employees as percentage of total work force, 1985.

1948–1976, scale economies accounted for something like 18% of net economic growth in the USA.

For the other OECD countries, especially Japan and West Germany, manufacturing constitutes a significantly larger share of GNP than it does in the USA (*Figure 7.5*). In the case of Japan, Denison attributed 44.4% of Japanese growth during 1953–1971 to scale effects; during 1970–1980 the relative importance of this factor actually increased to 54% according to the Japan Economic Research Center (JERC) (Usui, 1988, Table B-6). Accordingly, it is not unreasonable to suppose that economies of scale (i.e., the Salter mechanism) have probably accounted for a significantly larger share of postwar economic growth in the OECD as a whole, than in the USA – perhaps 40%. Finally, in the centrally planned CMEA countries, economies of scale and standardization have been exploited to an extreme degree and the share of employment in manufacturing has been even greater (*Figure 7.6*). It is very likely that a major technical reason for the recent economic

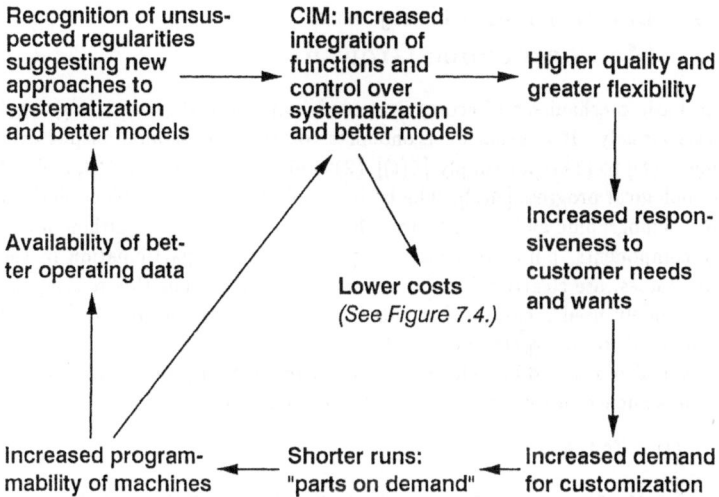

**Figure 7.7.** The new paradigm of economic growth. A possible chain-reaction mechanism to explain the transition to increased computerization and CIM.

difficulties of those countries is that the economic benefits of scale have been largely exhausted and replaced by diseconomies.

A major hypothesis arising out of the IIASA CIM project is that the Salter mechanism (*Figure 7.4*) may be obsolescent as a result of the increased emphasis on product variety and customization. A question of some interest is whether there is an alternative growth mechanism, illustrated schematically in *Figure 7.7*.

The key to the suggested "new" paradigm for economic growth is that increasing flexibility progressively reduces the cost differential between customized and standardized products. The smaller this differential, the greater the demand for diversity and, hence, flexibility. But this process, in turn, leads to further improvements in the manufacturing process, generating savings in both labor and capital and – in effect – restarting the traditional cost-driven engine of growth.

## 7.4   Labor Productivity and Macroeconomic Growth

The basic mechanisms of economic growth have been discussed in general terms already. It is usual for economists to ascribe economic output over time $[Y(t)]$ to (1) labor supply $[L(t)]$, (2) capital investment $[K(t)]$, and (3) technological progress $[A(t)]$. The labor supply is *exogenous*, being derived from demographic and social factors that change very slowly (although two key components, population growth rates and female participation in the labor forces, are clearly related to standard of living). For this reason, the most widely used measure of economic growth is total output per unit of labor input, or *aggregate productivity*.

It is also usual to introduce a simplified *production function* to describe the dependence of total output $Y$ on the three factors:

$$Y(t) = f(A, K, L) \ , \tag{7.2}$$

where $f$ is assumed to be a homogeneous function of the first degree in aggregate capital invested $(K)$ and aggregate labor supply $(L)$. Taking advantage of the assumed homogeneity property and dividing by $L$, yields an expression for the labor productivity $y$, viz.,

$$y = f(a, k) \ , \tag{7.3}$$

where the reduced variables are defined as:

$$y = Y/L \tag{7.4}$$

$$a = A/L \tag{7.5}$$

$$k = K/L \ . \tag{7.6}$$

Given this formulation one can, in principle, forecast future economic output in terms of the three factors $A$, $K$, and $L$; equivalently, one can forecast labor productivity $(y)$ in terms of the two ratios $A/L$ and $K/L$.

The forecasting problem can be simplified further for a closed economy by assuming the quantity of capital $[K(t)]$, a function of the gross capital in the previous year $[K(t-1)]$, gross output in the year, and two parameters (the depreciation rate, $d$, and the savings rate, $s$), i.e.,

$$K(t) = K(t-1)(1-d) + sY(t-1) \ , \tag{7.7}$$

where the product $sY(t-1)$ is the aggregate savings set aside (in year $t-1$) for new investment. Dividing by $L$ one obtains a convenient, reduced form,

$$
\begin{aligned}
k(t) &= k(t-1)\left(1 - \frac{sL}{L} + \cdots\right)(1-d) + sY(t-1) \\
&\simeq k(t-1)(1-d) + sY(t-1) \ .
\end{aligned}
\tag{7.8}
$$

Substituting (7.4) for (7.8) we obtain the following:

$$
k(t) \simeq k(t-1)(1-d) + sf[a(t-1), k(t-1)] \ ,
\tag{7.9}
$$

which is solvable iteratively given an explicit production function $f$ (such as a Cobb–Douglas or CES form), an explicit function $a(t)$, and starting values for $a$ and $k$, at some time $(t = o)$.

This simple basic growth model can be extended to the case of open economies by introducing an additional factor of production to reflect the import of intermediate goods (Krelle, 1989). Sectoral disaggregation can also be introduced (ibid). However, from our present (global) perspective, these extensions are conceptually straightforward, in principle, and offer difficulties mainly in practical implementation.

The last step is to decompose the function $a(t)$ into its parts. First, express $a(t)$ in terms of the annual *rate* of change $\dot{a}(t)$, namely,

$$
a(t) = \dot{a}(t)\Delta t \ ,
\tag{7.10}
$$

which has the convenient property that if $a(t)$ is an exponential (as usually assumed without further ado), then $\dot{a}$ is constant. It is easy to verify that for small annual rates of labor force growth $\mu$,

$$
\dot{a} = \frac{\dot{A}}{L}(1 - \mu + \mu^2 \cdots) = \frac{\dot{A}}{L(1+\mu)} \ .
\tag{7.11}
$$

The aggregate rate of technology change for the economy as a whole may, in turn, be decomposed by sectors, viz.,

$$
\dot{A}(t) = \sum_i \lambda_i \dot{A}_i \ ,
\tag{7.12}
$$

where $\lambda_i$ is the output share of the $i$th sector and $\dot{A}_i$ is the annual rate of technical change for the $i$th sector. It has been shown by Krelle (1985) that for the special case of a Cobb–Douglas production function the rate of change of technical change $A(t)$ in equilibrium is equal to the rate of change of labor productivity $\dot{y}(t)$, viz.,

**Table 7.1.** Technology levels and productivities.

| Level | Description | Relative labor productivity |
|---|---|---|
| 0 | Traditional | $y_0 = 1$ |
| 1 | Use of stand-alone NC and CNC machines | $y_1 = 2y_0$ |
| 2 | Use of FMCs and FMSs | $y_2 = 3y_1 = 6y_0$ |
| 3 | Complete integration (CIM) | $y_3 = 5y_2 = 30y_0$ |

$$\frac{dA}{dt} = \frac{dy}{dt} \ . \qquad\qquad\qquad\qquad\qquad\qquad\qquad (7.13)$$

Clearly the sectoral decomposition (7.12) is still applicable. Thus, one can now reverse the process and *construct* growth scenarios for the whole economy by assuming a given rate of technological change for the manufacturing sector or – just for the metal-working sectors – and assuming a "standard" rate of change (extrapolated from past history) for the other sectors.

The following scenarios were constructed to explore the macroeconomic effects. Output in each of the five metal-working sectors (381–385) is assumed to be allocated among four levels of technology, as shown in *Table 7.1*. The relative productivity levels correspond to data from a study of the Beretta Company (Italy) compiled by Jaikumar (1989a).

*Table 7.2* characterizes 11 countries (6 Western and 5 Eastern) in terms of the penetration in levels 1–3 in 1984 and (assumed) by 2000. Results of a simulation using a multi-country, multi-sector version of a simple growth model described above are shown in *Table 7.3* in terms of a comparison between model solutions "with" and "without" the accelerated introduction of CIM. The first column for each year displays "standard" growth vs. no technological progress; the second column displays the *differential* growth effect with and without (accelerated) CIM penetration. The third column cumulates this effect over time. It indicates, for instance, that with (accelerated) CIM total output in 1999 would be 5.6% higher than without CIM.

The economic growth simulation results cited in this last section were entirely based on a scenario in which labor productivity grows faster than "normal" in the metal-working sectors, while everything else remains the same. In particular, no change was assumed in the productivity of capital.

In the simple model described above *capital productivity* does not appear explicitly, so there is no way to change it. However, the effect on the model of increasing the productivity of capital is the same as the effect of increasing

**Table 7.2.** CIM diffusion scenarios: Shares in sectoral value added by levels of automation, percentage.

| Country | Level | 1984 Sector | | | | | 2000 Sector | | | | |
|---|---|---|---|---|---|---|---|---|---|---|---|
| | | 381 | 382 | 383 | 384 | 385 | 381 | 382 | 383 | 384 | 385 |
| USA | 1 | 6 | 13 | 10 | 13 | 9 | 28 | 47 | 42 | 57 | 42 |
| | 2 | 1 | 3 | 2 | 3 | 1 | 6 | 16 | 10 | 16 | 10 |
| | 3 | 0 | 0 | 0 | 0 | 0 | 1 | 2 | 3 | 2 | 3 |
| FRG | 1 | 4 | 14 | 8 | 11 | 9 | 31 | 48 | 45 | 58 | 45 |
| | 2 | 1 | 3 | 2 | 3 | 1 | 6 | 18 | 10 | 18 | 10 |
| | 3 | 0 | 0 | 0 | 0 | 0 | 1 | 2 | 32 | 3 | |
| Japan | 1 | 12 | 18 | 17 | 14 | 13 | 31 | 48 | 45 | 58 | 45 |
| | 2 | 1 | 4 | 2 | 4 | 1 | 8 | 20 | 12 | 20 | 12 |
| | 3 | 0 | 0 | 0 | 0 | 0 | 1 | 2 | 3 | 2 | 30 |
| France | 1 | 7 | 11 | 11 | 13 | 7 | 28 | 47 | 42 | 55 | 44 |
| | 2 | 1 | 3 | 2 | 3 | 1 | 6 | 16 | 10 | 18 | 8 |
| | 3 | 0 | 0 | 0 | 0 | 0 | 1 | 2 | 3 | 2 | 3 |
| UK | 1 | 7 | 11 | 11 | 13 | 7 | 28 | 47 | 42 | 55 | 44 |
| | 2 | 1 | 3 | 2 | 3 | 1 | 6 | 16 | 10 | 18 | 8 |
| | 3 | 0 | 0 | 0 | 0 | 0 | 1 | 2 | 3 | 2 | 3 |
| Italy | 1 | 7 | 11 | 11 | 13 | 7 | 28 | 47 | 42 | 55 | 44 |
| | 2 | 1 | 3 | 2 | 3 | 1 | 6 | 16 | 10 | 18 | 8 |
| | 3 | 0 | 0 | 0 | 0 | 0 | 1 | 2 | 3 | 2 | 3 |
| USSR | 1 | 3 | 6 | 5 | 6 | 4 | 12 | 25 | 20 | 25 | 25 |
| | 2 | 0 | 1 | 1 | 1 | 0 | 5 | 13 | 8 | 13 | 6 |
| | 3 | 0 | 0 | 0 | 0 | 0 | 0 | 1 | 2 | 1 | 1 |
| Bulgaria | 1 | 4 | 8 | 7 | 8 | 5 | 15 | 30 | 25 | 30 | 20 |
| | 2 | 0 | 1 | 1 | 1 | 0 | 5 | 16 | 10 | 16 | 8 |
| | 3 | 0 | 0 | 0 | 0 | 0 | 0 | 1 | 2 | 2 | 1 |
| CSFR | 1 | 4 | 8 | 7 | 8 | 5 | 15 | 30 | 25 | 30 | 20 |
| | 2 | 0 | 1 | 1 | 1 | 0 | 5 | 16 | 10 | 16 | 8 |
| | 3 | 0 | 0 | 0 | 0 | 0 | 0 | 1 | 2 | 2 | 1 |
| GDR | 1 | 4 | 8 | 7 | 8 | 5 | 15 | 30 | 25 | 30 | 20 |
| | 2 | 0 | 1 | 1 | 1 | 0 | 5 | 16 | 10 | 16 | 8 |
| | 3 | 0 | 0 | 0 | 0 | 0 | 0 | 1 | 2 | 2 | 1 |
| Hungary | 1 | 4 | 6 | 5 | 6 | 4 | 15 | 25 | 20 | 25 | 15 |
| | 2 | 0 | 1 | 1 | 1 | 0 | 5 | 13 | 8 | 13 | 6 |
| | 3 | 0 | 0 | 0 | 0 | 0 | 0 | 1 | 2 | 1 | 2 |

Source: Dobrinsky, 1990 (see Volume IV).

**Table 7.3.** Estimates of the "pure" macroeconomic effect of the CIM diffusion: Results from *ex ante* simulation runs.

| Country | 1990 | | | 1995 | | | 1999 | | |
|---------|------|------|------|------|------|------|------|------|------|
|         | $\Delta^a$ | $\Delta_{84}^b$ | $\Sigma\Delta_{84}^c$ | $\Delta^a$ | $\Delta_{84}^b$ | $\Sigma\Delta_{84}^c$ | $\Delta^a$ | $\Delta_{84}^b$ | $\Sigma\Delta_{84}^c$ |
| USA     | 0.37 | 0.41 | 2.01 | 0.33 | 0.40 | 3.99 | 0.30 | 0.38 | 5.59 |
| FRG     | 0.55 | 0.57 | 2.66 | 0.60 | 0.65 | 5.14 | 0.61 | 0.69 | 7.85 |
| Japan   | 0.47 | 0.52 | 2.44 | 0.50 | 0.59 | 5.28 | 0.50 | 0.64 | 7.76 |
| France  | 0.49 | 0.50 | 2.38 | 0.51 | 0.53 | 5.01 | 0.51 | 0.53 | 7.14 |
| UK      | 0.30 | 0.30 | 1.44 | 0.32 | 0.33 | 3.04 | 0.33 | 0.34 | 4.38 |
| Italy   | 0.34 | 0.36 | 1.74 | 0.35 | 0.38 | 3.62 | 0.34 | 0.39 | 5.15 |
| USSR    | 0.31 | 0.31 | 1.31 | 0.35 | 0.40 | 3.17 | 0.38 | 0.47 | 5.35 |
| Bulgaria | 0.36 | 0.43 | 1.88 | 0.39 | 0.53 | 4.34 | 0.39 | 0.55 | 6.50 |
| CSFR    | 0.60 | 0.74 | 3.22 | 0.68 | 0.92 | 7.51 | 0.67 | 0.98 | 11.34 |
| GDR     | 0.51 | 0.59 | 2.47 | 0.60 | 0.78 | 6.03 | 0.63 | 0.89 | 9.43 |
| Hungary | 0.29 | 0.33 | 1.40 | 0.36 | 0.43 | 3.41 | 0.39 | 0.48 | 5.31 |

[a] $\Delta = \frac{Y^{CIM}-Y^o}{Y^o} \cdot 100\%$, where $Y^{CIM}$ is GDP (NMP) in the model solution with CIM diffusion; $Y^o$ is GDP in the model solution with zero rate of growth of technical progress.
[b] $\Delta_{84} = \frac{Y^{CIM}-Y^{84}}{Y^{84}} \cdot 100\%$, where $Y^{84}$ is GDP (NMP) in 1984.
[c] $\Sigma\Delta_{84}^T = \sum_{t=1984}^{T} \Delta_{84}$.
Source: Dobrinsky, 1990 (see Volume IV). Source: Ayres (1984), data from US Department of Labor.

the *amount* of capital (i.e., the capital stock) with constant productivity. This can be done in the scenario by tinkering with the constants in equation (7.7). There are two possibilities, to *increase* the savings rate, $s$, or to *decrease* the depreciation rate, $d$. (In reality, the depreciation rate is more likely to *increase*, since machinery that is used more intensively will wear out sooner.)

Additional growth scenarios, as well as other intersectoral and international comparisons, will be discussed in Volume IV.

## 7.5   Product Mix: The Implications of Diversity

It is tempting to discuss economic benefits of new flexible technology in manufacturing as though product mix were not an issue. Yet with unit cost differences of 100-fold, or more, depending on the scale of production, the product mix of the economy as a whole (and changes therein) become crucial considerations. An obvious strategy for cutting costs and increasing productivity throughout the economy would be to (somehow) rearrange things so as to produce more things on a very large scale and fewer things on a very small sale. An all-powerful central planner, well aware of the power of economies of scale, might theoretically intervene to achieve such a shift, with great social benefits. (It may well be the case that central planners in CMEA countries have attempted to intervene in this way.) But there are inherent limits.

The manufacturing sector must, in the very nature of things, produce at all scales from the lowest on up. The reason for this has already been set forth, at least implicitly: manufactured goods are themselves produced by machines and equipment that are products of the manufacturing sectors. In this context, it is noteworthy that the capital goods sector is, and must be, capable of reproducing itself. But components are made on a larger scale than the products they go into. Similarly, final goods destined for consumers are made on a larger scale than the machines that make them, which are made on a larger scale than the underlying infrastructures, and so on. Some critical items, such as prototype designs, will probably always be unique.

In practice, some consumer products are manufactured on an extremely large scale, viz., hundreds of thousands per year (watches, cameras, consumer electronics, personal computers, bicycles, cars, sewing machines, household appliances, handguns, ammunition, hand tools, and so on). Components common to a wide range of products, such as ball bearings, threaded fasteners, electrical connectors, microchips, and small electric motors, are manufactured on an even larger scale (millions or tens of millions). Goods manufactured on this scale are characteristically produced in dedicated plants, which make one item only or at most a family of minor variations (models) of an item. However, as will be pointed out subsequently, the demand for variety is nowadays increasing faster than the demand for quantity.

Many capital goods (and military goods) are manufactured on an intermediate scale, in the range 1,000–100,000. Examples include minicomputers; trucks and buses; small aircraft; aircraft engines; outboard engines; diesel engines; boats; farm equipment; standardized machine tools; forklifts; commercial refrigeration, air-conditioning, dry cleaning, cooking, baking, dishwashing, and laundry equipment; office copiers, typewriters, and switchboards; commercial video and movie cameras and projectors, sound equipment, etc.

On a smaller scale of production, made in small numbers, from 10 or less to perhaps 1,000 one comes to commercial and military aircraft, rockets, and missiles; railroad locomotives; subway cars; gas turbines; large diesel engines; elevators; escalators; transfer machines; specialized types of machine tools; large electric motors; large pumps; mainframe computers; central-office telephone-switching equipment; heavy-duty off-road vehicles for construction and mining; tanks and armored vehicles; radar systems; navigation systems; sonar systems; electric generators, and so on.

Manufactured products in the one-time-only or "very few" category include dedicated manufacturing systems (discussed above), large buildings (including heating, lighting, and ventilating systems), ships, metro systems, spacecraft, and the like. In this category, also, are the prototype models, industrial patterns, jigs and fixtures, and so on which are an important part of every factory (and which, in fact, provide the main source of work for most small machine shops).

It is clear that items in the capital goods category are produced in far smaller numbers than consumer items. On the other hand, because of the very pronounced economies of scale noted above, capital goods account for a comparatively large fraction of the total value of manufactured products, as shown in *Figure 7.8*. This figure applies to the USA. It probably underestimates the fraction of capital goods in the output of Japan, West Germany, East Germany, CSFR, and the USSR. For less advanced economies, on the other hand, the fraction of capital goods is likely to be smaller than it is in the USA.

Given the potential benefits of modularization and standardization of capital goods indicated in the example of automobile engines discussed previously, the following implication is inescapable: If flexible, general purpose tooling could be standardized (or built up from standard modules) and manufactured in large enough numbers to capture some of the economies of scale characteristic of the metal-working industry, then the (fixed) capital costs of manufacturing can be cut very sharply. The relative impact on unit costs of final products is proportionately larger, the smaller the number of items being produced, and largest of all on things like factories.

To be sure, it is not realistic to think of capital costs over the whole economy dropping by a factor of 10. But, over a period of years – before the year 2000 – it is entirely realistic to envision cuts of the order of 10 in some cases and 2-fold or 3-fold overall. This gain is attributable partly to *economies of scale* in the production of capital equipment, and partly to *economies of scope* in the manufacturing process of consumer goods. These gains are apart from, and in addition to, any savings from other sources (such as inventory reductions) that may result from the use of advanced manufacturing technologies in the capital goods manufacturing sector.

It is almost redundant to point out that the essential technological prerequisite of modular general purpose machine tools is flexible design and

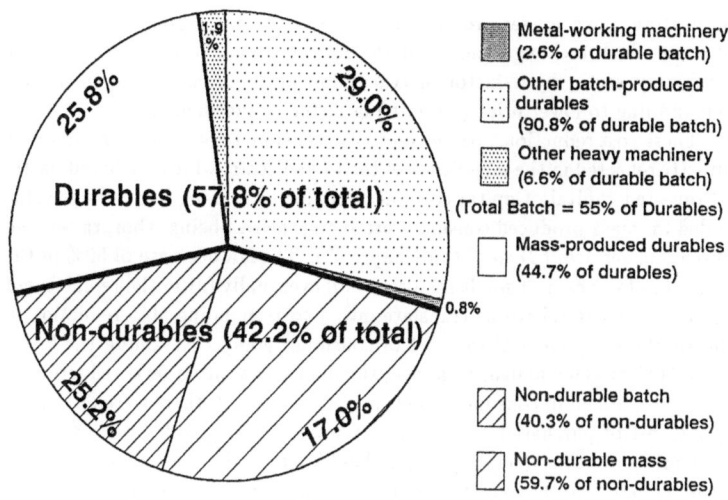

**Figure 7.8.** Distribution of manufacturing value-added (V.A.). Total V.A. = $585,000,000 (1977 $). Source: Miller, 1983.

computer control (CNC). Similarly, the benefits of reducing design and engineering costs through the use of computers (CAD/CAE) are evident. Thus, the economic benefits of modularity, discussed above, can only be achieved by means of computerized automation, i.e., CIM.

Another implication of *Figure 7.8* (taken together with *Figure 3.4*) is that – other factors remaining equal – technological innovations leading to reductions in the unit cost of capital goods, especially machinery, could result in a sharp decrease in the absolute size of the capital goods sectors while simultaneously lowering the costs (but not necessarily the size) of the consumer goods sectors.

Let us take a specific scenario to make the argument clear. Suppose the average unit cost of production machinery were cut by a factor of three from its present level. The rest of the capital goods sector would presumably enjoy somewhat lower costs for the same reasons. This assumes a combination of modularization and standardization of the hardware component, and reprogrammability of the software component to provide flexibility, thus permitting much longer production runs and the corresponding economies of scale. Such a change might cut the total revenues of the metal-working machinery sector by a factor of two (allowing for modest growth in unit demand due to lower prices, permitting faster replacement).

These cost reductions, passed on to the consumer goods sectors, would be directly reflected in the magnitude of their fixed costs, which are based partly on invested capital. Capital costs account for something like 30% of value added for mass-produced consumer products, the rest being labor, taxes, and purchased services ( *Table 3.1* in Chapter 3). Thus, a decrease of 50% in the revenues of the capital goods sectors would eventually (after the replacement of the whole capital stock) translate into a corresponding 50% reduction in the *capital component* of the cost of consumer goods. Assuming the latter to be 30% of value added, as above, this translates into a 50% reduction in total value added in consumer goods manufacturing. If there were no change in the cost of purchased materials, this would make possible a 7.5% cut in final prices to consumers. But purchased materials must also eventually decline in cost, to the extent that they also share in the capital cost savings. Because each sector's input of purchased materials is the final output of another sector, costs and prices of all manufactured goods on the average would probably decline by an amount somewhere in the neighborhood of 10% (assuming land and building costs were unaffected).

Such a price cut, in turn, would have some impact on final demand for consumer goods, via the Salter mechanism ( *Figure 7.4*). As noted previously, price elasticities of demand range from 0.3 to 0.7 in most sectors. Given lower real prices, consumer expenditure for manufactured products would increase, with corresponding increases in the revenues of the manufacturing sectors of the order of 3% to 7%.

The growth impulse from the Salter mechanism as applied to capital goods depends on the actual numerical value of the average price elasticity, $e$, as noted in the Section 7.2. For the OECD countries, where most consumption goods markets are approaching saturation, the net benefits are unlikely to be great in magnitude. For the developing countries and the CMEA countries, on the other hand, markets are far from saturated and a major growth impulse can be expected.

This discussion considers the economic impact of new manufacturing technology on costs, prices, and final demand, assuming no changes in product mix (variety). This is an unreasonable and unrealistic assumption, however. It "puts the cart before the horse," in some sense, because it neglects the strong likelihood that the new, computer-based manufacturing technology is being introduced largely as a response to sharp increases in the variety (and complexity) of products demanded by consumers. (The growing importance of complexity was noted earlier in Chapter 1.) If we take this evidence seriously, then the economic impact of the new technology must be assessed in terms of how manufacturers respond to these trends.

Given this perspective, the primary question is: How is the product mix of the economy changing? The logical second question one might ask is: To what extent will new manufacturing technology affect product variety? When these questions are answered satisfactorily, at least to first order, one can logically address the third question, namely, What are the implications for costs and demand?

But no sooner is the second question posed than an answer of sorts suggests itself. Surely, the introduction of new advanced manufacturing technologies *per se* is likely to have little direct effect on the product mix of the economy. Why should it? Asking the question this way puts the cart before the horse. The use of computers in manufacturing will not cause changes in the product mix, except insofar as the manufacture of computers and controls themselves contributes to that mix. The product mix of the economy is changing, certainly. But in all probability this is occurring for other, more fundamental reasons. So we are back to the primary question: How is the product mix of the economy evolving, for reasons other than the introduction of computers into the manufacturing process?

Here we must depart from the realm of standard economics into somewhat uncharted realms, since the conventional core assumption of neoclassical economics is that all prices reflect an equilibrium. If that were so, it would mean that nothing could or would change except due to exogenous factors, which by definition lie beyond the realm of economics. A complete discussion is provided in Ranta *et al.* (forthcoming, Volume V).

However, for the present it may suffice to hark back to the discussion in Chapter 1, where it was noted that the demand for product variety and customization seems to be increasing quite rapidly. This is what one would expect, given a long period of rising incomes and saturation of demand for basic services. Ford's classic experience with the Model T (and Volkswagen's similar experience with the Beetle) seems to be repeating on a broader front throughout much, if not most, of the manufacturing sector. In the automobile sector, by far the fastest growing segment is the "luxury" category, where almost no two units are identical in all respects. A typical manufacturer offers several thousand variants. For example, Audi's assembly plant produces an exactly identical car, on the average, only once in 20,000 cars

(Warnecke, 1989). In effect, more and more people are "trading up" from low-priced standardized products to more elaborate versions with many options in design and performance, catering to individual tastes.

## 7.6   Productivity Slowdown

This is an area where there are nearly as many theories as theorists. A point of departure is the well-documented slowdown in productivity growth since 1973–1974, as shown in *Table 1.2* in Chapter 1.

Explanations that have been put forward range from the "end-of-catchup" hypothesis (Abramowitz, 1989) the "energy crisis" hypothesis (Jorgenson, 1988), the "deceptive measurement" hypothesis (Griliches, 1988, 1989), and the "new paradigm" hypothesis (Freeman, 1989). All probably reflect a part of the truth, which makes it all the harder to make defensible generalizations about the subject. Nevertheless, this section outlines still another perspective that deserves attention.[1]

Having emphasized the increasing role of information processing in manufacturing already, the outlines of a simple hypothesis are fairly obvious. As long as humans retain their exclusive role as "information transducers," they will be potential bottlenecks in any macro-level information processing system. But, whereas machines typically reveal their overloaded state in some straightforward way (or can be designed to do so), humans and organizations do not. Indeed, there may be no simple test for being overloaded.

What appears to be happening in many traditional organizations (firms) is this: with the help of computers, individual office workers performing particular information-processing functions have become more productive. They generate more quantity and better quality outputs than ever before. But data entry, data verification, error detection, and error correction continue to be manual processes, for the most part. Data have to be moved from file to file, often by semi-manual methods (via tapes or floppy disks). Information "generated" by a computer usually ends up in a pile of printouts on someone's desk. This point has often been made before. For instance, Martin Baily (1986) has argued that the "office revolution" has largely created more paperwork.

The problem is that the computerized output of one office is often the data input of another – not necessarily administratively related. As the output of one office increases, the load on some other office increases at the same time. Increasing local information output (productivity), may inadvertently overload the downstream consumers of that information, not individually but organizationally. Errors, in particular, can cause organizational overload and external costs.

Two examples may illustrate this point. In the 1950s a number of mail-order book clubs were created, usually offering a choice of several best-sellers at low prices in exchange for a promise to buy several more books during the year on a regular basis. The idea was that the books would be shipped automatically unless the customer cancelled the order in advance. Of course, there were inevitably cases where the communication went awry and the customer got a book he or she did not want. In some cases the customer would keep it and pay for it, but in others he or she would send it back. But anybody who joined such a club will remember the number of letters or phone calls needed to straighten out the account after such an incident. Many book clubs collapsed under the weight of unanticipated clerical work needed to unscramble computer-generated confusion. The airline "frequent traveller" clubs of recent years offer comparable opportunities for mistakes, and also required a comparatively large amount of human effort to correct them.

In short, the result of increased productivity in one place can be confusion, delay, and an apparent need for more manpower or more computer power elsewhere in the organization to perform information-processing tasks that might be better not done at all.

The analog here is the cardinal sin of machine-shop management: it is to "optimize" the operation by encouraging each steward to maximize the utilization of his or her machines and operators. This means keeping them working as much of the time as possible, if only to make parts for later use (i.e., inventory). The standard argument for this is that "they will be needed someday, if not now." This strategy of optimizing at the machine level is disastrous at the plant level, because if demand increases it invariably results in overloading and delays at some critical "bottleneck" operation, which is being kept busy making items that are not needed while interfering with production of those that are needed.[2]

In short, the information-processing activities in large organizations are comparable, in a sense, to manufacturing in a job shop. Somewhere in every information-processing organization (at any given time), there is a critical human bottleneck. More likely than not, the bottleneck has to do with correcting errors or mistakes originating elsewhere in the system. When this (human) bottleneck is overloaded, the organization has production problems, most likely in terms of delivery delays or quality control or both.

In summary, it is plausible that it is the indirect effects of coping with problems of overloading and interference created by the availability of too much information of the wrong kind that is responsible for the failure of local increases in productivity to be translated into global productivity growth. If this view is on target, the slowdown in productivity growth is occurring not in spite of the information-processing power of computers, but actually because of it. The fact that the effect has been most noticeable in the USA, where computer usage spread most rapidly, is consistent with this hypothesis.

If this view is correct, the problem will eventually be resolved when – and only when – computers handling operational information are able to "talk" to each other directly, bypassing many lower-level clerks and functionaries who are currently responsible for feeding information into computers and interpreting it. It is primarily the errors introduced by such clerical personnel, and the extensive efforts needed to find and correct these errors, that results in the overload problems discussed above.

## Notes

[1] This section is largely based on Ayres (1989).
[2] For a readable nonmathematical exposition of the critical difference between local and global optimization in manufacturing, see E.M. Goldratt and J. Cox, *The Goal*, N. River Press, 1984.

# Chapter 8

# Social Impacts of CIM

## 8.1  Introduction

Unquestionably, some of the most immediate effects of the advent of computer integrated manufacturing or CIM will be the impacts on working conditions, skill requirements, and employment opportunities. Relevant aspects of the problem include labor displacement, structural changes in the labor force, changes in work content, and changes in the work environment. The importance of these problems is reflected in a considerable number of papers and books (see, for example, Engelberger, 1980; Haustein and Maier, 1981, 1985; Hunt and Hunt, 1983; Ayres and Miller, 1983; Nilsson, 1984; USOTA, 1984; Howell, 1985; Leontief and Duchin, 1986; Ayres, 1988b).

There is now convincing evidence that, at the micro-level (i.e., within the firm), the application of CIM (including robots) is accompanied by direct labor displacement and changes in skill requirements. Many semiskilled clerical, draftsman, materials-handling, machine operator, and assembly jobs are gradually being eliminated. A somewhat smaller number of highly skilled jobs is being created at the same time. Some of these new jobs are in software development and engineering, and some are for multi-skilled technicians who will supervise the flexible manufacturing systems (FMCs, FMSs, and LS/FMSs).

There is also general agreement upon the fact that the current state of CIM application has not yet (as of 1990) led to measurable changes in the employment level or employment structure (e.g., the qualification structure) at the national level. This is hardly surprising, in view of the adoption/diffusion lags (Chapter 5).

**Table 8.1.** The development of white-collar workers in a large multinational car manufacturer.

| | Index numbers showing trends | | | | | Actual % of white-collar workers engaged in activity, 1987 |
|---|---|---|---|---|---|---|
| | 1983 | 1984 | 1985 | 1986 | 1987 | |
| Organizational change | n.a. | n.a. | 100 | 118 | 136 | 0.5 |
| Innovative activities and product development | 100 | 124 | 150 | 153 | 145 | 35 |
| Portfolio management, banking and insurance and risk reduction | 100 | 116 | 106 | 98 | 102 | 1 |
| Factory production | 100 | 102 | 107 | 113 | 118 | 49 |
| Marketing and sales | 100 | 104 | 109 | 112 | 112 | 6 |
| Education | n.a. | n.a. | n.a. | n.a. | n.a. | n.a. |
| Welfare tasks | 100 | 115 | 107 | 119 | 98 | 2 |
| Coordination | 100 | 108 | 108 | 126 | 126 | 1 |
| Procurement | 100 | 128 | 145 | 153 | 168 | 5 |

Opinions differ more with regard to the medium- and long-term employment impacts of CIM. Some authors emphasize the problem of job replacement accompanied by higher unemployment. Other authors are of the opinion that the labor-saving effects of CIM application will actually lead to higher productivity and income, resulting in higher domestic demand (and improved export competitiveness) ultimately creating a net increase in total employment.

The best analogy is with agriculture. Even at the beginning of the twentieth century most Americans and Europeans lived on farms and grew their own food. They sold the surplus to the cities, of course, but farming was the largest single occupation. Today, as we enter the last decade of the twentieth century, the percentage of workers needed for food production is of the order of 5% or less. Approximately a quarter of all workers in the OECD countries (under 20% in the USA) are employed in manufacturing. I think this will drop to 10% by the middle of the next century.

On the other hand, even now only two-thirds of the workers employed by manufacturing are classified in occupations involving direct contact with what is being produced, e.g., laborers, truck drivers, machine operators, assembly workers, welders, inspectors, and so on. One-third of the workers are only indirectly involved with making the product. They are office workers, stock clerks, maintenance personnel, sales personnel, programmers, drafts-men, engineers, supervisory personnel, and executives (see *Table 8.1*). Both groups will be cut back in numbers by CIM, but humans will still be needed for the off-line jobs, including design, engineering and prototyping, main-tenance and repair of the production machines, and all of the higher-level functions. The jobs eliminated will be machine operative jobs and "infor-mation transducer" jobs such as data entry and typing reports.

# 8.2 Quality of Work Life

Factory work in the past has often been unpleasant, sometimes extremely so. Noise, dirt, long hours, danger, and boredom all contributed. But "de-humanization" seems to have been the greatest problem for workers, and one that becomes more severe the more tightly organized and "Taylorized" the manufacturing system. Robert Southey, a poet and commentator in the early nineteenth century, contrasted the traditional agrarian society with the emerging factory system and noted the latter to be

> a system which...debases all who are engaged in it...a system in which the means are so bad, that any result would be dearly purchased at such an expense in human misery and degradation, and the end so fearful, that the worst calamities which society has hitherto endured may be deemed light in comparison with it (Southey, 1829, Tierney *et al.*, 1967).

Another nineteenth century critic, John Ruskin, wrote:

> We have much studied and...perfected, of late, the great civilized invention of the division of labour; only we give it a false name. It is not, truly speaking, the labour that is divided, but the men: – Divided into mere segments of men – broken into small fragments and crumbs of life: so that all the little piece of intelligence that is left in a man is not enough to make a pin, or a nail, but exhausts itself in making the point of a pin or the head of a nail.... (Ruskin, 1851, cited by Östberg, 1986).

The dehumanization theme has been repeated many times since then. Among the many luminaries who have condemned "the machine" more or less com-prehensively were Thorstein Veblen and Lewis Mumford. Perhaps the most enduring image of the early twentieth-century factory environment is that of Charlie Chaplin, the worker who cannot control the machine, but is con-trolled by it (*Modern Times*). For a century or more the factory environment has been comprehensively criticized as "inhuman," a cause of increasing alienation, a cause of the decline of the traditional work ethic, and so on.

This theme is still very much alive in the industrial sociology and "human factors" literature. Since the 1950s this literature can fairly accurately be described as anti-Tayloristic in substance and in tone.[1] Taylorism (or neo-Taylorism) has become, in effect, the principal target of the labor movement, in both Europe and the USA.

The main point is that workers basically do not like working in factories, no matter how "ergonomically designed." The bigger the factory, the less they like it. The more routine and humdrum the job, the less they like it. The brighter the worker, the less he or she will enjoy working in a factory environment. The very qualities that make a worker eligible for promotion, in most organizations, will make him or her dislike the repetition that characterizes Taylorized manufacturing systems. An apocryphal quotation by Nelson Algren (1944) describing conditions on the assembly line runs as follows:

> I tell you boys, when you put on a wheel here and a roller there and a belt in the other place, it ain't long till you get to be hell on wheels and no brakes, and it was goodbye crapping a smoke or drinking a rest. If you had to hold up two fingers like a kid in school, you'd meet yourself coming back or you'd know the reason why. You had to pick 'em up and lay 'em down right there at your post and make believe you like it or ask them to pull your card. You could just walk outside talking to yourself if you couldn't stand the gaff (Algren, 1944).

The Taylor system of (so-called) scientific management treated workers as if they were passive, interchangeable parts in a machine for manufacturing. Specialists analyzed each task and specified not only the task, but the exact motions to be performed. All of this was coordinated and controlled by a hierarchical, many-level management structure and a rigid system of work rules. (Not surprisingly, detailed job specifications and work rules were later used by labor unions as a mechanism for job proliferation and preservation. The sins of the fathers are visited on the sons.) No wonder factory workers felt dehumanized.

Careful attention to so-called human factors in the work place – cheerful colors, pleasant background music, clean locker rooms, coffee breaks, quality circles, and so on – can improve the situation somewhat. But the factory (no matter how disguised) will never be selected as a place of work by anyone with a reasonable alternative. The factory has been a kind of way station for refugees from rural poverty, or immigrants.

This is a fact of life. It is therefore not surprising that Henry Ford chose to (in fact, had to) pay higher wages than his competitors to attract good workers to his efficient but dreary assembly lines. There is, *ceteris paribus*, a trade-off between wages and working conditions. Factories do have to pay more than offices, for example, to attract workers of equal caliber. This point will have relevance later.

The dissatisfaction of workers in traditional Taylorized work environments has been a worry for many years, since the 1940s at least. Philip Herbst of the Tavistock Institute of Human Relations in the UK was one of the pioneers in exploring new "socio-technical" forms of organization. The key idea was what Herbst called "minimum critical specification" of jobs. In effect, workers should be treated as adults and allowed to decide on their own how a job can best be done, based on experience with actual conditions.

However, the Scandinavians have carried the Tavistock ideas further. Einar Thorsrud of the Work Psychology Institute in Norway led the way toward practical implementation, using autonomous groups (teams). Swedish social scientists such as Bertil Gardell and Reine Hansson became interested. In 1969 three Swedish organizations, the Swedish Employers Confederation, the Confederation of Trade Unions, and the Central Organization of Salaried Employees took up the idea and formed a joint organization that, in turn, created a task force to sponsor research projects. Meanwhile, industry moved ahead without waiting. SAAB's engine assembly division introduced the autonomous group system to auto production in 1972. In 1974 Volvo applied the idea to its new automobile assembly plant. The Atlas–Copco Mining Co., another Swedish firm, also adopted the idea during the early 1970s. The results of this experiment were reported in *Scientific American* and have sparked wide interest (Bjork, 1975).

Since then the movement has spread increasingly, especially in smaller firms in Sweden and Finland. Both Volvo and SAAB have also had good results, although only (so far) in new plants. The autonomous group system has worked particularly well in plants where new flexible manufacturing systems (FMS) and just-in-time (JIT) systems are being introduced (e.g., Bouchut and Besson, 1983). It has even been argued by some of the leading experts in management science that the new approach to organizing work, which blurs the "white-collar" vs. "blue-collar" distribution, is, in fact, essential to the successful implementation of new FMS and other sophisticated hardware (Bessant and Haywood, 1988; Hayes and Jaikumar, 1988). This conclusion is strongly supported by our own case studies, reported in Ranta *et al.* (forthcoming, Volume V).

Thus, for those still employed in factories (at least in the industrialized countries) in future decades, the quality of work life is likely to improve significantly.

## 8.3   Network Organization

While not in any sense a direct implication of CIM, it is worth noting that the anti-Taylorist trend away from vertical integration *within* organizations has a possible analogy at the next level, i.e., in terms of changing relationships between and among organizations. To make this notion clearer, consider the familiar distinction between the "nuclear" family (two generations, parents and minor children, living in a single household) and the multi-generational "extended" family that is traditional in many other societies, especially China.

US firms are generally more similar to the "nuclear family." There are large conglomerates, such as GE or Teledyne, with many relatively independent subsidiary businesses, to be sure. However, US banks are legally excluded from "active" stock ownership (since 1933), and in most conglomerates the components are totally owned subsidiaries: they have no independence at all. By the same token, boundaries *between* firms have tended to be quite high and impervious. This is partly due to the long history of antitrust laws in the USA, which has strongly discouraged informal cooperation with long-term suppliers or direct competitors, for instance.

Japanese business groups (Keiretsu) typically resemble extended families. They consist of a bank, a trading company (Sogo Shosha), one or more "flagship" manufacturing companies, and a number of satellite firms and suppliers. They are all linked financially by cross-ownership of stock (often controlled by the group's bank) and by long-term purchase/supply relationships. Early retirees from the flagship firms often spend their later years with suppliers, or even with new startups under the protective umbrella of the group.

The European industries use both organizational styles. Many large West German firms, for instance, are actively controlled by one of the big banks (Deutsche Bank, Commerz Bank, and Dresdner Bank). Many large European firms (especially in France and Italy) are still nationalized or semi-nationalized, with significant levels of government ownership. And national flagships in key industries (Volvo in Sweden, Fiat in Italy, Renault in France) are still common in Europe. Long-term supply relationships are also quite normal.

It would be natural to conclude from the relative economic success of Japanese and West German industry in recent years that long-term cooperation among firms is more beneficial than the more competitive inter-firm relationships characteristic of the USA. Such a conclusion may well be warranted, although the evidence is perhaps not yet unambiguous.

One interesting possibility for escaping some of the disadvantages of the current US pattern of industrial organization is that so-called value-added partnerships (VAPs) will grow. The VAP is a new organizational form that intensively uses modern communications and information processing technology, but harks back to the "putting out system" of organizing cottage industries that preceded the first industrial revolution.

It has been argued by economic historians that the economies of scale (and division of labor) favored the vertically integrated factory over the earlier system. If so, then the approaching decline of scale economies as a dominant factor driving economic growth (Section 7.3) may tilt the balance toward other network organizations. In particular, the VAP is "a set of independent companies that work closely together to manage the flow of goods and services along the entire value-added chain" (Johnston and Lawrence, 1988).

The best-known example of a VAP in the USA is McKesson Corp., a \$6.7 billion drug and health-care enterprise. By offering independent, owner-operated drugstores access to a sophisticated computerized system that they could not afford individually, and which helped them compete successfully against the chains, McKesson also helped itself become a major player in the market.

But the trend to VAPs really began in the Italian textile industry clustered around Prato. In the early 1970s a few vertically integrated textile giants dominated the industry, and they were in trouble. The first to convert itself to a VAP was the family-owned firm of Menichetti. Starting in 1972, Massimo Menichetti split the firm into eight independent organizations and began to sell 30%–50% of the stock in each of its key employees – to be purchased out of profits!

> After only five years, Menichetti's productive unit had over 90% utilization of their machines. Both labor and machine productivity had increased. New machines had been added, increasing capacity by 25%. Product variety increased in each of the eight units from an average of 600 to 6000 different

yarns. Average in-process and finished-goods inventory dropped from four months to 15 days (Johnston and Lawrence, 1988).

This innovation was so successful that it was imitated. Today, 15,000–20,000 independent small businesses – linked both horizontally and vertically by computerized information nets and cross-ownership – have replaced all except one of the Italian textile giants. And, from 1970 to 1982, while European textile output declined sharply, production in the Prato area more than doubled.

Indeed, the Japanese trade companies are, in effect, VAPs. Book publishing has evolved in this direction. And the US auto industry is taking tentative steps away from vertical integration. Chrysler has created a VAP with its suppliers, distributors, and labor union. Ford, too, has been moving in this direction. (Significantly, the world's most successful auto producer, Toyota, produces internally only 20% of the total value of its product, as compared with 30% for Chrysler, 50% for Ford, and 75% for GM.)[2]

Of course, the Italian textile model may not be directly applied to other industries. Moreover, it is clear that "networking" *per se* is not a sufficient condition for success. Enforcement of "rules of behavior" among members of the VAP, to discourage predatory behavior of various kinds, is essential. But, having said this, a future industrial structure consisting of VAPs or relatively small integrated plants with specialized capabilities is more plausible than a revival of mass production.

## 8.4   Employment and Unemployment

It is clear that one motivation for introducing some of the elements of automation and CIM – notably robots – is the opportunity to reduce direct labor costs (Chapter 5). This means that some net labor displacement must be expected. At the first level of analysis, the only serious questions are: Who is most likely to be displaced and how fast will the displacement occur?

The clearest (and best-documented) displacement is the substitution of robots for semiskilled machine loaders, operators, and assemblers. As noted in Chapter 5, Japan has led the way in this regard for several reasons. In terms of the present discussion, the main point is that industrial production has been rising much more rapidly in Japan than in other industrial countries. This has led to a labor shortage in Japan (which, for cultural reasons, has been very reluctant to adopt the German policy of importing "guest workers").

The lack of unemployment in Japan makes the problem of reabsorbing displaced workers relatively easy to handle. Also, as is well known, the large Japanese firms are comparatively paternalistic in their treatment of workers. In effect, the major Japanese firms regard all male workers up to age 55 as "lifetime" employees. They are treated for corporate accounting purposes as fixed costs rather than variable costs (Chapter 3). As a result, workers do not fear the introduction of new technologies. In fact, they typically welcome machines taking over dull, dangerous, or repetitive tasks.

The other major reason for Japanese preeminence in robotics also stems from the rapid increase in industrial production in Japan. This has resulted in many new "greenfield" factories being built from the ground up. A new plant is more likely to use advanced manufacturing equipment (such as robots and FMSs) than an older one. Not only has this massive investment created an enormous domestic market for production machinery and capital goods. It has also emphasized the the latest generation of equipment, as opposed to replacements and upgrades of earlier generations of equipment. This has been particularly beneficial to the Japanese robot and CNC machine tool producers, such as Fanuc and Yamazaki. (It is in sharp contrast to the situation recently confronting US machine-tool producers who have had to rely much more on cyclic factors, such as the massive investment in "down-sizing" US-built cars in the late 1970s. (This generated a temporary "bubble" of demand, followed by sharp cutbacks in the early 1980s.)

The so-called displacement ratio (workers per robot) has been variously estimated in the literature, depending on application and period. The Japanese data shown in Chapter 5 are probably the best that have been compiled anywhere. The data suggest that in the early years (when the high cost of robots permitted only the most promising applications) around three workers were displaced by each robot. Currently in Japan, the ratio is closer to unity. In the other major industrial countries, where the use of robots lags by five to six years, as compared with Japan, the displacement ratio is still probably somewhat higher (perhaps 1.5). However, for purposes of extrapolation, the best estimate is probably that each robot will displace one job.

If the automobile industry is reasonably typical, one might expect that about 40% of all persons now employed in mass production can and will eventually be displaced by a robot. As far as robots are concerned, this is probably as good an estimate as anybody can make at present. It goes without saying that many other semiskilled jobs will also disappear, being displaced (in effect) by computer-intensive technologies from CAD/CAM to GT and FMS.

One way to estimate the employment displacement effect is to turn the expected labor productivity improvements upside down. We want to compare two hypothetical scenarios, namely, (1) the "standard" scenario in which labor productivity in each sector continues to grow at its historical rate and (2) the "CIM" scenario in which the labor productivity in the metal-working sector grows at an accelerated rate. For example, suppose the labor productivity in the metal-working sector as a whole for the CIM scenario is denoted $y_c$, as compared to $y_0$ for the standard scenario, while the proportion of total GNP (Y) produced by that sector is the same for both cases. If the labor used in the sector is $L_c$ and $L_0$, respectively for the two scenarios, it is easy to show that

$L_c/L_0 = y_0/y_c$ .

The ratio of the sectoral labor requirements in the two cases is inverse to the ratio of labor productivities for the sector. Hence, if the CIM scenario is characterized by a labor productivity double that of the standard scenario, the sectoral labor requirements will be only half as great as they otherwise would be. In this hypothetical case, if the sector would employ ten million workers in the standard scenario, it would only employ five million in the CIM scenario. (As a matter of interest, the data in *Table 7.2* is enough to compute the labor productivity ratio for the sectors 381–385 for each of the 11 countries.)

Given the output or employment share of each three-digit sector, an aggregated ratio for the entire metal-working sector can easily be generated. Assuming, for the sake of simplicity, that the entire metal-working industry adopts CIM at the same rate as the 383 sector, the productivity ratio for all six industrialized countries represented in *Table 7.2* would be roughly 0.45 + 0.42 × 2 + 0.1 × 6 + 0.03 × 30 = 2.79. In this case the labor ratio would be 1/2.79 = 0.36 (36%). This example is not intended to be realistic – it is almost certainly too extreme with regard to both productivity and labor impacts. It may be regarded as an upper limit. However, it calls attention to the close link between the two calculations. (More realistic economic scenarios are discussed in Volume IV.)

In comparison with the GNP as a whole and the labor force as a whole, the metal-working industry and its labor force is relatively small – even in the FRG and in the centrally planned economies of Eastern Europe where services are as yet undeveloped. In the US metal-working industry the difference between the standard case and the CIM case is unlikely to be greater than half or two-thirds of the level implied by the above example. Still, manufacturing employment in metal-working sectors could decline by three to four million by the year 2000, in the USA alone.

Luckily, the 1990s are a period during which the rate of new entrants to the work force will decline, relative the 1970s and early 1980s when postwar baby boomers came of age. This is most notably true for the USA. Given that all new jobs have been in the service sectors for quite a long time (since the late 1960s), a continuation of recent economic growth rates would virtually force a decline in manufacturing employment to accommodate projected labor requirements in the service sectors.

Of course, this is certainly an overoptimistic view. Attrition alone (at normal retirement ages) cannot be expected to cut the manufacturing work force by more than one-third over the next decade. Attrition accounted for only about 55% of the job losses due to successful introduction of robots in the UK prior to 1983 (Fleck, 1984). There will be more massive job cuts. Many will occur owing to plant closings where a large, old manufacturing facility becomes obsolete and not worth rehabilitating. For example, there have been a number of major plant closings as large, but declining, manufacturing firms – such as American Motors and International Harvester – have been forced to "restructure" radically. When this occurs, the plant closing is often in an old industrial city where there is little new investment.[3]

It is also a fact of life that – due to lack of education, age, and cultural rigidity – few displaced manufacturing workers are able to find equivalent alternative jobs, on their own, especially if they are unwilling or unable to move to another location. Many are eventually forced to take other employment as union benefits and savings run out, but typically at significantly lower wages. Some of these people are the "new poor." Many of these unemployed workers in the USA are children or grandchildren of immigrants from abroad or from poverty-stricken rural environments. Quite a few are blacks – the last to be hired and the first to be laid off when seniority is the criterion. For some, it is a case of from rural poverty to urban middle class and back to poverty in two generations. We return to this topic in the next section.

The narrow and quantitative issue of displacement in the metal-working sectors is not the whole story, of course, or even the most interesting part of it. (It is the only part we can begin to quantify, however.) In the first place, CIM technology is directly applicable throughout manufacturing, and indirectly it will have significant impacts on much of the service sector as well. In the second place, CIM is only one of the applications of Information Technology (IT). The spread of IT is having an effect on all aspects of society, and on employment patterns in particular. Christopher Freeman (1989) has suggested that IT is triggering the evolution of a "new techno-economic paradigm" characterized by information-intensive, flexible technology, as contrasted with the energy-intensive, mass- and flow-production technology earlier.

## 8.5    The Social Structure and the Role of Unions

One of the most persistent worries expressed by social critics, and one of the subtlest, is that the end of traditional manufacturing, with its highly paid (unionized) workers, means a drastic shrinkage of the middle class. Worse, it could follow that the traditional path for upward mobility, from the working class into the middle class, will be blocked. (Recall the point made earlier, that factory work has tended to be a refuge for poverty-stricken farm workers displaced from the countryside or new immigrants. This fact has been elevated into a kind of virtue by modern-day intellectuals of the labor movement.)

At any rate, the emergence of a two-tier labor structure, and the concomitant decline of the middle class, has been suggested by labor unionists and by journalist-authors like Barry Bluestone (*The Deindustrialization of America*), Bob Kuttner (*The Declining Middle*), and Studs Terkel (*The Great Divide*).

The basic argument runs along the following lines. There is a good deal of evidence (albeit, hardly conclusive, as yet) that the advent of computerized automation is polarizing the work force in terms of skills. That is, it is increasing the demand for highly skilled workers – especially computer technicians, programmers, and engineers – while eliminating many semiskilled machine-operator jobs. Historically, the latter workers have been quite well paid. Consequently, during recent decades blue-collar workers have constituted a significant part of the "middle income" group.

The following states the concern as seen from the perspective of organized labor in the USA:

> As computers and robots take over more and more functions in the factory and the office, a two-tier work force is developing. In some cases, jobs are being upgraded. In many other cases, jobs are being downgraded.

(In some production processes, computerization may lead to a narrowing of skill differentials between supervisors and production workers when both need detailed knowledge of a relatively complicated process.)

At the top will be a few executives, scientists and engineers, professionals, and managers, performing high-level, creative, highly paid, full-time jobs in a good work environment. And the executives among them will decide whether the work will be done by people or by robots, whether the work will be done in Terre Haute or Taiwan.

At the bottom will be low-paid workers performing relatively simple, low-skill, dull, routine, high-turnover jobs in a poor work environment. These jobs will often be part-time and usually lacking job security and opportunities for career advancement. Too often these jobs are over-supervised and lacking in any control over the pace of work.

Between these two major tiers will be fewer and fewer permanent, well-paid, full-time, skilled, semi-skilled and craft production and maintenance jobs which in the past have offered hope and opportunity and upward mobility to workers who start in low-paid entry-level jobs. Many middle management jobs will also be gone (AFL–CIO, 1983).

So much for fears with regard to the manufacturing sector itself. On the other hand, it is alleged that a service economy "employs legions of keypunchers, sales clerks, waiters, secretaries and cashiers, and the wages for these jobs tend to be relatively low" (Kuttner, 1983, p. 60).

In short, it is feared by some that the shift toward a service-based economy will leave most of the people who are no longer needed to operate punch presses or weld auto bodies in lower-wage jobs. This, in turn, would *ipso facto* increase the maldistribution of wage income. In addition, of course, labor-oriented critics of automation also argue that cuts in earnings by workers would have a counter-Keynesian depressing effect on the larger economy, by reducing their buying power and, hence, their contributions to aggregate demand. This topic was discussed in Chapter 7 in the context of macroeconomic impacts.

Two other distinct issues, here, tend to be confused in many discussions and need to be sorted out. One is the notion, which should be reformulated as an hypothesis, that factory jobs are inherently higher paid than service jobs. The other is that change in the income distribution in the direction of increased inequality is socially undesirable.

With regard to the second issue, the socially desirable distribution of income is a matter not of economics but of ethics and politics. Most of the individuals involved in this study would tend to the personal view that inequality is already excessive, at least in some countries (notably the USA), but this is a value judgment. Social choice must be left as a matter to be settled, ultimately, by a political process.

If current national patterns are assumed to reflect the conscious wishes of the electorates, one would have to conclude that Americans (along with a few other nationalities) are satisfied with an income distribution with a small professional and managerial elite receiving a disproportionately large income, while the lowest-paid tier of society is living at or below the "poverty line." This group has grown larger in the last decade. By contrast, most European countries and Japan have a much narrower range between the lowest and highest incomes (Scandinavians being the most egalitarian), while the countries of Eastern Europe permit virtually no differentials – at least in theory, if not in practice.

But these patterns have changed significantly over time, and are changing still. In the past the United States was less tolerant of great wealth differentials than it seems to be now, whereas most of Europe became far more egalitarian during the decades immediately after World War II. Today Eastern Europe and Scandinavia seem to be reversing that trend. In future decades it seems likely that some convergence will occur, with the USA moving to correct some of the current imbalances and Europe (especially in the East) moving toward a somewhat less regulated pattern and (probably) somewhat greater inequality.

With regard to the first issue – the hypothesis that factory labor is inherently better paid than service work – the main evidence is entirely empirical, with little or no theoretical basis. Factory jobs in the USA, Japan, and Europe are better paid, by a considerable margin, than many service occupations, especially in the retail sector. In addition, the most rapidly growing job categories in the industrial countries are relatively low paid, while the best paid (i.e., factory) jobs available to those without higher educational qualifications are generally declining in number. The percentage of total employment in the manufacturing sector has declined particularly rapidly in the 1980s (*Figures 8.1* and *8.2*), albeit West Germany and Japan have resisted the general trend.

Still, the question remains whether there is anything in the nature of factory work that attracts (or requires) higher compensation than service work at the same level of skill? (Actually, there is no truly satisfactory way to compare skill levels required for very different types of work, but we cannot pursue that fascinating topic here.) The most obvious – but not necessarily complete – answer is that factory work is inherently less pleasant than office work or service work, hence must command a higher wage to attract workers.

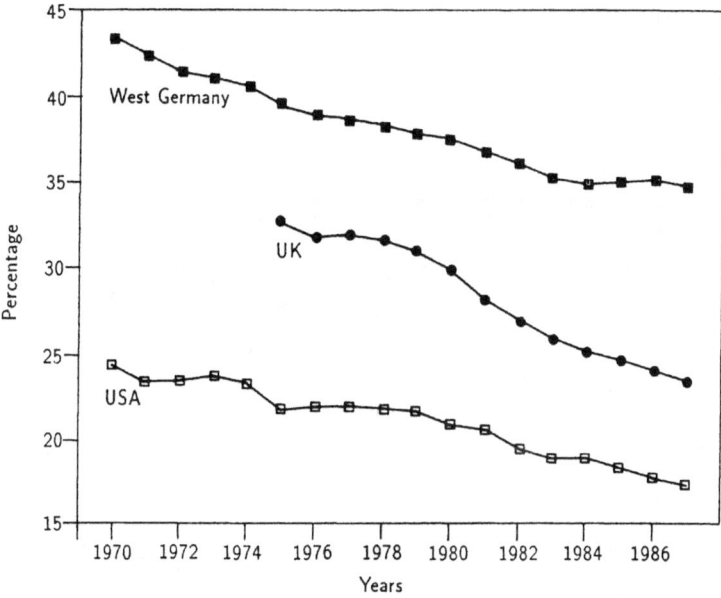

**Figure 8.1.** Percentage of total employment in manufacturing: West Germany, the UK, and the USA. Sources: West Germany, Statistisches Bundesamt Wiesbaden; UK, CSO National Accounts; USA, Dept. of Commerce.

There is a modicum of truth in the above argument, especially insofar as it explains historical patterns, as pointed out above.

Having said this, it must also be said that high wages in factories nowadays tend to reflect unionization. The factory, as it evolved in the nineteenth and twentieth century, was an ideal "hothouse" for organizing workers into unions (and political action groups). Most workers got no satisfaction from the work itself, because of its routine nature. Applications of crude Taylorism made the problems of boredom and *anomie* worse by reducing the worker's control over any aspect of his or her job. In every country the unions had to struggle very hard for legal recognition, and in some countries workers' organizations were repressed by law backed by police power and armed force. The predictable result was to unify their membership: repression helped

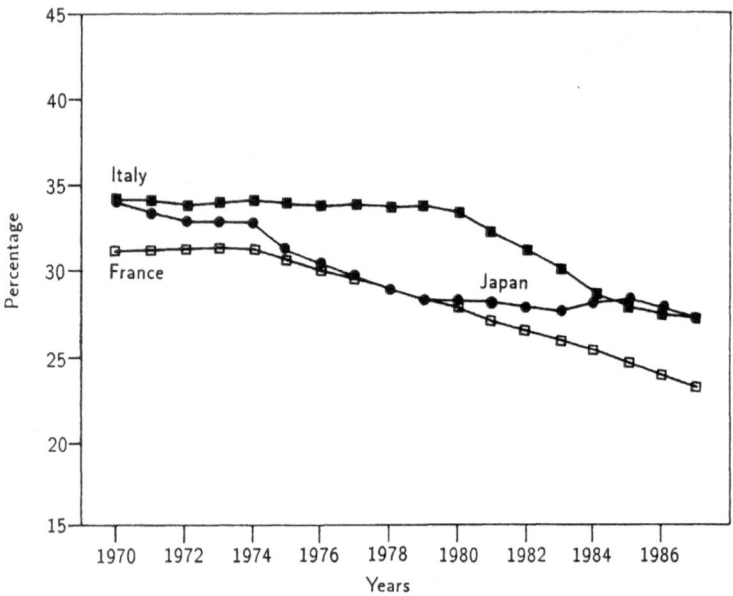

**Figure 8.2.** Percentage of total employment in manufacturing: Italy, Japan, and France. Sources: Italy, OECD; Japan, Economic Planning Agency; France, OECD.

workers achieve a considerable degree of solidarity. In most of Europe, the result was an extreme political polarization of Right vs. Left, with the Left dominated by Marxists.

In Europe the labor-oriented parties achieved considerable political power in the late nineteenth century, but the revolutionary rhetoric of the extreme Left and the Bolshevik revolution of 1917 led to a political reaction and a climate in which fascism flourished. Socialist but non-Communist labor parties regained their influence in Western Europe after World War II, while the Communists took power in Eastern Europe with the help of Soviet military force. In the USA the legal rights of labor unions were strongly resisted until the Roosevelt administration in the 1930s, when labor unions were given a legal exemption from the antitrust laws, which would otherwise have prohibited a labor monopoly. However, while unions were politically influential in the Democratic party from the mid-1930s until the 1980s, they were never able to dominate it completely because of the nonunion "southern wing" of the party.

In the USA, unions used their legalized monopoly power over labor, at least in the industrialized northeast and midwest, to achieve immediate and significant economic gains. During the postwar decades, especially, economic growth was seemingly effortless, and rising wages in large industrial sectors like steel and automobile manufacturing in North America and Europe could be paid for with first out-of-productivity gains achieved by simple economies of scale. Later, for a well-established industry with oligopolistic dominance in the national market, it was possible to continue to raise wages and pass the higher costs along to the public in terms of higher prices. As a result, by the 1970s the unionized sectors in North America and Europe were receiving wages anywhere from 50% to 100% above the wages of workers (sometimes, even workers doing the same jobs) in nonunion firms or sectors.

One exception to this pattern is Japan, where industrial and craft (occupation) unions on the Western pattern do not exist. In fact, a major confrontation between management and unions occurred in the early 1950s in Japan, and the unions were defeated decisively. As a result of that defeat, workers' organizations in Japan are, essentially "company unions," i.e., employee organizations. Having no monopoly power on the larger scene, Japanese unions cannot use a strike as an effective weapon and have become essentially captives of management. This may have been a case of very good fortune in disguise!

The other obvious exception is the group of Communist (or former Communist) countries belonging to CMEA, once known as the Eastern bloc. The USSR has had a Communist government since 1917, and the other countries in the group were ruled by Communist governments from the late 1940s until 1989. Only in the case of Poland was a non-Communist labor union (Solidarity) legalized before 1989, and the Polish case does not provide a clear basis for extrapolation to the other countries. In some sense, the Eastern countries, until recently, constituted a merger between (Marxist) labor unions and the government itself.

The disparity between wages in union and nonunion sectors in Western Europe and North America can probably no longer be sustained in a competitive international environment. It can only be sustained (if at all) by political means at the national and supranational level. Some European countries, especially in Scandinavia, have virtually no nonunion workers.

This "social contract" keeps wages (and taxes) very high, but tends to induce firms – when they are free to do so – to move their operations to other countries or regions where unions are less powerful and labor costs are lower. This tendency is real, in both Europe and North America, even though labor costs are no longer the dominant factor in total costs. If they were, the "offshore" tendency would be much more pronounced than it is.

During most of the twentieth century, as production processes have become more capital intensive, the financial threat of interruptions to production has grown apace. As long as firms cannot produce and earn income without their unionized workers, they are vulnerable to strikes. This is still the situation in virtually the entire manufacturing realm. However, further automation will unquestionably reduce the financial threat of strikes.

The privately owned telephone systems in North America illustrate the point. Unionized workers perform all installation and maintenance tasks, but since the advent of automatic switching equipment several decades ago, they do not perform most of the normal on-line call-switching functions. Thus, if the union calls a strike, the telephone system remains in operation, with only minor degradation of service. Supervisory employees carry out essential repairs and maintenance. Lacking power to cause crippling financial losses, the telephone employees' unions have been unable to push wages far above competitive levels – and strikes have been rare and short.

One long-range implication of CIM is that workers will eventually play a similar role in manufacturing. That is, they will perform off-line functions of all kinds, from design and marketing to after-sales service. They will design and maintain the machines that do the actual manufacturing. They will do customizing and prototyping. But, for significant periods of time, most manufacturing processes will be able to function with supervisory personnel only, i.e., without the unionized on-line workers. This is no pipe-dream fantasy. Some Japanese factories (e.g., Fanuc and Yamazaki) already operate unmanned during the third shift, and do so routinely.

When a firm reaches this stage of automation, it can no longer be shut down for long periods by its unionized workers. (Approximately 200 plants, worldwide, will be at this stage by 1991 or so.) It will be in a position to resist wage claims that are not warranted by productivity improvements. This ability to keep costs down will, eventually, force its foreign competitors to follow suit, unionized or not. Japanese competition has already forced US-based automakers to close many plants and eliminate hundreds of thousands of jobs. The power of labor unions to control wages and working conditions for its membership is waning, even as more enlightened modes of work organization are spreading.

The bottom line can be stated succinctly: In a competitive, nonunion (or company union) world, factory workers would tend to command somewhat higher wages than workers with similar qualifications in other jobs. The differential cannot be determined precisely, but it is probably greater than 10% but less than 50% (a good guess might be 20%). It is noteworthy that manufacturing workers in the USA receive an average premium of 50% over service workers (mostly nonunion), whereas the premium for factory work is 18% in West Germany and 8% in Japan.

On the other hand, in some industries and some countries, factory workers have become used to being paid a good deal more than they could expect to receive in a competitive labor market. Moreover, for competitive reasons, these are just the industries most likely to introduce advanced automation at the earliest possible opportunity. (A GE vice president once characterized the alternatives as "automate, emigrate, or evaporate.") Hence, the days when the middle class included large numbers of relatively uneducated and semiskilled but highly paid factory workers protected by union membership are already history. Unions need not disappear, but their power to inflate wages beyond their natural level has already largely disappeared.[4]

To return to the question with which this section began: Is the middle class declining? The most plausible answer seems to be *yes*, in the short run, although for the longer run the answer is probably *no*. As noted already above, the unskilled factory job was once seen as a painless route into the middle class ("The American Dream"). This was an anomaly. In any case, it is history – a thing of the past. Such jobs will not disappear overnight, of course, but new entrants into the labor force in the industrialized countries will have almost no opportunity to get such jobs – except (for a while) in Japan. Moreover, the few such jobs that remain will lead nowhere except to early displacement and unemployment.

## 8.6   Education

In the longer run, society will travel a new road to social advancement. Actually, it is an old road, which was very effective in the nineteenth century (but which has been neglected recently in the USA, especially). That road lies through the school system, the community colleges, the technical institutes, and the universities. It might be termed: *lifetime education*.

Every study of the implementation of flexible automation and CIM emphasizes the need for training – and retraining – workers to use the new technologies. To state the case in a simple way, flexible manufacturing requires flexible workers. Skills to operate particular types of equipment, for instance, will always be important, of course. But no skill will last a lifetime in a fast-moving world. It is the ability to learn new skills in a reasonably short period of time that is needed.

This ability, or its lack, differentiates the educated from the uneducated. Education provides learning, to be sure, but far more important are the tools for learning. The first of them, which is more attributable to the home environment than the schools, is *language*. The ability to use and understand verbal communication is the most fundamental tool, for without it the others are largely foreclosed. Anyone who doubts this should consider the enormous problems encountered by people unfortunate enough to be born deaf in trying to learn to speak. It is through verbal communication, at the beginning, that children learn how to express ideas in words.

The second important tool, obviously, is the use of written language – *literacy*. Under modern conditions, there is almost no job in industry that does not require at least basic literacy. A truck driver must, for instance, be able to understand maps, traffic signs, and labels – at the very minimum. The lowliest stock clerk must have the same ability. It would be suicidal to try to operate complex machinery without being able to read instruction manuals and warning signs. Even a gardener must know enough to distinguish fertilizer from insecticide.

But these most fundamental levels of learning are far from enough to qualify a worker for any occupation that will command a wage above the lowest level in the future. *Numeracy*, the ability to understand numbers and perform basic arithmetic operations, is almost as important as literacy. It is absolutely essential, for instance, for any retail sales clerk or waiter. Such a person may well later use an electronic calculator for increased speed and accuracy, but a person who does not understand the basic operations cannot ever be trusted to use a calculator or cash register correctly.

But basic literacy and numeracy are no longer adequate for the purpose of performing any responsible or truly skilled job. In the Taylorized factory of half a century ago, most workers were basically employed for their muscles and hand-to-eye coordination. (In a very real sense, they were like programmable robots.) In the flexible factory of the future, there will not be many workers, but those who are there will be computer literate. They will also be flexible enough to do many jobs, and to learn quickly to do new jobs as required. They will also have to teach others.

This requires an in-depth knowledge of the machinery's purpose and how it works. The worker on the factory floor need not know enough about engineering to design the machine, but he or she will need to know roughly what the machine designer was trying to achieve and what can go wrong. The worker will also need to know what the product designer and the sales department expect, and what the factory manager needs. Likewise, each of those people also must know a good deal about the other's job.

A report based on the International Symposium on Microelectronics held in Japan in 1985 summarized the capabilities required of workers: "adaptability to rapid changes in the facilities and in job contents; the capacity to acquire knowledge and techniques on electronics and control systems; and the problem-solving ability to meet the requirement of quality improvement" (*Report ISML*, 1985, p. 42). The same report also emphasized "*continuous* education and vocational training throughout the workers' career to help them, improve their adaptability to technological changes" (ibid, p. 43, italics added).

In short, the specific knowledge that is needed on the job cannot, in general, be acquired anywhere except on the job. But, since the job keeps changing, a core part of the job is continuous learning itself. The skills of learning to learn effectively are not achieved in eight or nine years of basic schooling. In fact, they are achieved in large part as a by-product of learning other things.[5] It may be possible, in the future, to teach people to be better and more efficient learners. But, for the present, the most effective way is simply to learn more – to stay in school longer, and to return to the classroom often throughout working life.

It is obvious that the specific knowledge that most people learn in school, even in college, contributes very little to later occupational requirements. The accountant has no need for physics or French literature; the architect needs to know nothing of geography or social science, and so on. Yet it is also clear that education correlates quite well with income. This correlation would be even more striking if the scholarly community – people who later become teachers and professional researchers – were omitted. Individual exceptions and caveats are easily cited, but the overall conclusion is unavoidable: education is what gives workers greater value to employers and, consequently, greater wage-earning capacity.

What sort of education? Beyond the obvious need for training in the use of computers there is much more to be said. This question is clearly of the utmost importance, and it must be a major item on the domestic policy agenda of every industrialized or industrializing country over the next decade. It cannot be answered here except to note that flexibility and creativity may well be culturally determined. If so, it is important not only to teach the "right" subjects (such as communication skills, mathematics, and computer literacy) but also to avoid teaching the "wrong" things or the wrong way. Problem-solving ability and self-confidence are unlikely to be developed, for instance, in an educational system that stresses memorization of facts.

The other major question that must be addressed by policymakers in the future is how to make educational services of adequate quality available to the lower tier of the social system. The likelihood of a significant back-sliding of some families from the middle class into poverty seems clear. It does not follow automatically that the "decline of the middle" is *necessarily* permanent. It is not.

However, many sociologists have noted the *poverty syndrome* starting with poor living conditions, poor schools, and early dropout, leading to unemployment and resulting in continued poverty. This syndrome is common, but not inescapable. However, people who relapse back to poverty are likely to be demoralized at best, and most will not escape from the trap without help. Only the government can provide the necessary transitional assistance. However, to discuss the details of social policy in this field is beyond the scope of this book.

## 8.7    International Trade and Economic Development

The CIM study as it has been conducted to date has not systematically addressed the impact of CIM diffusion on the development strategy of the newly industrialized countries (NICs), such as Taiwan, Korea, and Brazil, still less on the Third World. It is to be hoped that such a study will be carried out in the near future, as a follow-up and complement to the present one.

As a starting point for any such analysis, one must recognize that the present international distribution of income is extremely skewed, with a few countries enjoying extremely high standards of living, while billions of people live in very poor circumstances. Moreover, whereas two decades ago there was some confidence that the gap would close itself, with some assistance from the industrialized countries, the last decade has essentially been a retrogression. One might list many reasons for this, but three among them are worth serious consideration here:

(1) *The debt crisis.* The Third World is awash with unpayable debt. This largely dates back to the surplus cash in Western banks in the late 1970s that, in turn, was created by large cash flows accruing to OPEC countries after the "oil shock" of 1973–1974 and the second round of high oil prices in 1979–1980. Since then, of course, petroleum prices have sharply declined in real terms, many former exporters (such as Mexico, Nigeria, and Venezuela) are now in deep financial difficulties, having overspent their windfall wealth. The debt is a huge burden to the Third World, generally, and a threat to the world banking system.

(2) *The balance-of-payments crisis.* Until 1980–1981 there was a long-term increasing net inflow of capital to the Third World from the advanced industrial countries. This money was used to finance purchases of capital equipment (and other imports) from the advanced countries. Bank debt grew each year, but it was assumed that growing GNPs would justify this (as the US government assumes that its own internal debt will not have to be repaid, as long as the economy grows enough to cover the interest costs). However, as the oil price has dropped since 1980 the banks have had less surplus cash to loan and are much less anxious to loan to countries perceived to be on the brink of insolvency. Without money to finance economic growth, net repayment of existing debt has been demanded by the banks, even as the conditions for repayment have become more and more difficult in almost every developing country. The direction of capital flow has now been reversed. With large net repayments to the industrialized countries ($50 billion last year, up $12 billion from the year before), the only possible way to balance the deficits in the financial accounts is to earn a surplus from exports of manufactured goods or raw materials.

(3) *The export imperative.* Some newly developing countries (NDCs) with export-oriented industrial sectors have done well in recent years. But Latin America, Africa, and South Asia – countries with weak manufacturing sectors, a history of import substitution, and high domestic tariffs – have, in general, the heaviest debt loads. They cannot compete internationally and therefore cannot balance their financial accounts by exporting manufactured goods into the markets of the OECD countries. These countries are therefore in the most severe difficulties. Such

countries (unless they can attract tourists on a massive scale) basically have three choices: (i) to sell raw materials or agricultural products, (ii) to sell assets, or (iii) to declare bankruptcy and refuse to pay their debts. The only assets of interest to potential buyers in the developed world are likely to be raw-material-based assets (land, mineral rights, etc.). Bankruptcy would probably mean no international credit for ten or more years, which would make financing needed purchases of capital goods and spare parts even more difficult than it is already.

The existing situation is clearly a vicious circle. Poor countries are forced to sell raw materials (or, more or less equivalently, to develop raw-materials-based industries) to pay the interest on their past development loans. They cannot sell most manufactured goods competitively because their industries are too small and technologically obsolete.[6] They cannot modernize their industries because they cannot borrow any money to do so, and their limited hard-currency earnings are needed for debt service. *Their desperation to sell raw materials in a glutted market drives market prices lower and lower and makes the problem ever worse.*

It is obvious that the existing situation is intolerable for several reasons. One reason is that it is already breeding political extremism and instability. In environmentally fragile parts of world, such as the Sahel and the tropical rain forests, the life-support system is breaking down with increasing frequency. The consequences are famine, pestilence, and waves of sick and destitute refugees. Not many find their way to the rich lands of the north, but enough do to create serious social and political stresses. Even sending them back to their poverty-stricken homelands – as the "boat people" are currently being repatriated from Hong Kong to Viet Nam – causes acute political pain.

Another reason why the situation is intolerable is that the dependence of the poor countries on exporting raw materials is holding down the prices and increasing the usage of those materials – especially fossil fuels. This, in turn, is causing environmental problems whose scope and seriousness we are just beginning to grasp. The industrialized world is beginning to debate ways and means to reduce environmental pollution from coal burning, for example. At the same time, developing nations like China and India are planning enormous increases in energy consumption and other activities that harm the environment, to accelerate their own economic development. The conflict in objectives is clear (WCED, 1987).

To resolve the conflict and restart the development process that has been stalled for a number of years, a new economic development strategy is needed. Clearly it must be export oriented, since the developing world as a whole is not self-sufficient in either goods or technology. To be sure, many products that are currently imported from the north could, in principle, be imported from other countries in the south. Development in south–south trade is surely one element in the new approach.

But south–south trade will not suffice, especially in the realm of sophisticated capital goods, such as airliners, computer systems, telecommunication systems, and advanced production equipment. These must be imported from the north. Obviously some of these imports can and will be paid for with raw-materials exports (after all, even with enhanced materials recycling and more intensive energy conservation, the north will not be self-sufficient in raw materials, especially petroleum).

However, relatively few of the developing countries can depend on raw-materials exports to finance development, and (most of those are in the Middle East). For the rest, there is really no alternative to developing export-oriented manufacturing industries. The once-popular protectionist policy of import substitution and subsidy of "infant industries" (still widely practiced, especially in Latin America) has proved to be seriously flawed. As implemented in most places, it has created inefficient, high-cost, non-competitive firms – often nationalized – that would be devastated by the cold wind of serious competition, even from other countries in the south. Protectionism at the national border is quite simply counterproductive.

What can be done? The political and macroeconomic aspects of the problem are complex and beyond our scope. Such issues as the feasibility of free south–south trade, while maintaining some protection for the south against the rougher competition from the north, may have to be explored. (And how could this be done without central planning of some sort?)

One statement can be made with a good deal of assurance, however: Manufacturing industries in the developing countries need CIM as much as, or even more than, their counterparts in the north. Protectionism for south–south trade, at least, must be phased out and large multinational firms specializing in different fields must be allowed to evolve. This means that the intensity of competition will increase far more in the south, in coming decades, than it has in the markets of the industrial world (which were already fairly competitive). It follows that quality and cost, rather than job protection, are going to be dominant considerations in corporate decision making.

The problem of quality control and managing complexity exists everywhere. It is only less severe in the south to the extent that the products manufactured there are themselves less complex and sophisticated. CIM will be needed in the developing world for precisely the same reasons it is now being adopted in the developed countries.

## Notes

[1] A recent book on the subject by Rose (1978, p. 31) begins its chapter on "Taylorism" as follows:

> It is difficult to discuss the "contribution" of F.W. Taylor to the systematic study of industrial behavior in an even-tempered way. The sheer silliness from a modern perspective of many of his ideas, and the barbarities they led to when applied in industry, encourage ridicule and denunciation.

[2] Peter Drucker, for instance, has characterized the Japanese form of industrial organization as a "flotilla," in contrast to the traditional (vertically integrated) US "battleship" (Drucker, 1990). He states: "The factory of 1999 will be an information network."

[3] Examples are easily identified. Northern West Germany is suffering significant unemployment at present, while the south (Bavaria and Baden-Württemberg) enjoys boom conditions. The same pattern exists in the UK, France, and the USA.

[4] Defining the "natural" level of wages for manufacturing could be an arcane and controversial topic. No two economists or social scientists are likely to agree on the precise criteria that should be employed in such a determination. But wages in a (mostly) nonunion but enlightened manufacturing environment such as the US electronics industry would be a reasonable indicator. The natural level at present is probably around $10–$12 per hour (including benefits) for a semiskilled individual such as a machine operator.

[5] "Other things" may include geography, literature, languages, science, cooking, or basket weaving. It probably does not matter much. In any case, there are other reasons for learning, besides developing the ability to learn more effectively and the flexibility that results.

[6] The obvious exception to this statement is handicrafts. Such goods *are* exportable (if the quality is high) and future prices will probably rise. Those countries with significant reservoirs of skilled hand-workers (e.g., carpet-weavers, cabinetmakers, etc.) may be able to develop further in this area.

# Epilog

## Post-Industrialism Reconsidered

The term "post-industrial society" was popularized among intellectuals in the mid-1960s with the publication of the American Academy of Sciences' *Toward the Year 2000: Work in Progress* (1968) and Daniel Bell's *The Coming of the Post-Industrial Society: A Venture in Social Forecasting* (1973). These books were quickly followed by several popular books, such as Alvin Toffler's *Future Shock* and *The Third Wave*, which created a widespread impression that the world is entering an "age of information" that will make factories and physical production processes obsolete. Nothing could be further from the truth.

Of course the growing importance of information in symbolic form is undeniable. Equally undeniable is the rapidly growing relative importance of information-intensive activities, in terms of the numbers of persons employed in various occupations. But it does not follow that information-intensive activities are in any sense a substitute for materials-intensive activities. To be sure there are some cases where substitution does (or may) occur, for instance, the possibility of telecommuting to work from a work station in the home and thereby saving gasoline and reducing tire wear.

However, on closer scrutiny, it is not easy to identify many instances where the use of information can actually replace the use of materials, except insofar as increased materials and energy efficiency results from more sophisticated technology requiring more information processing, etc. Indeed, the relationship is complementary. Sophisticated information technology, which depends on silicon chips engraved with highly complex patterns of electronic pathways, requires very sophisticated factories. Yet relatively few humans may work in these factories, because of the dangers of contamination and because human workers are too clumsy and too error-prone to perform the critical operations. Machines do the actual manufacturing work. Humans are there in small numbers to supervise, to diagnose problems when things go wrong, and to fix the machines when necessary. Humans also design the chips, the fabrication equipment, and the tests.

It is sometimes said that information flies around the world "at the speed of light." This may be an exaggeration, but the speed at which information travels has increased in recent decades. In the nineteenth century the telegraph was a revolutionary invention because it permitted messages to be sent across continents or oceans in a few hours, as compared with the days or weeks that were previously required. But compared with the information technologies of today, the telegraph was primitive. Photographs, diagrams, computer programs, and formulae can literally travel at the speed of light; even physical copies of products can be shipped across the globe in a few hours.

One clear implication of this communications speedup is that there is little advantage in being an innovator, at least for certain types of products: it is easier and cheaper to be an imitator. Thus, the low-wage workers in Hong Kong, Taiwan, South Korea, and elsewhere on the "Rim of Asia" have become known for turning out cheap, somewhat inferior copies of brand-name products originally designed and produced in the USA or Western Europe. A few decades ago Japan did the same thing and had the same reputation.

But wait a moment. Does this imply that manufacturing is unimportant? Exactly the contrary. Without sophisticated manufacturing capabilities those East Asian imitators would be no better off than the Third World countries. Moreover, while the *design* of a consumer product such as a radio or a camera may be relatively easy to reproduce in another location, *this is not true of the manufacturing technology itself*. If it were true, every country would possess advanced manufacturing technology, and there would be no poor countries!

With regard to the USA, it is worthwhile to recall some thoughts recently put forward by Lester Thurow, dean of MIT's Sloan School in the *New York Times* (September 5, 1989), under the headline "The Passing of America's Post-Industrial Era." Thurow argues that the slow growth in labor productivity in the USA since the 1970s is due to low capital investment combined with rapid growth in demand; ergo, many new low-wage service jobs were created. If the USA had enjoyed West Germany's rate of productivity growth, from 1972 to 1983, it would have added only 3.5 million new jobs compared with 14.4 million actually added.

But now Thurow argues that the pendulum in the USA is about to swing once again. Assuming that the persistent US trade deficit *cannot* continue indefinitely, consumption must fall or domestic manufacturing must eventually increase once again to take up the slack. This will require increased investment in labor-saving (CIM) technology in the factories. It will also raise wages in the service sector. In Thurow's view, this process is inevitable. The only unresolved question is the ownership of the production facilities. We need not concern ourselves with that issue.

In fact, in apparent recognition of the seriousness of this problem for the developing countries, the Ministry of International Trade and Industry (MITI) of Japan is planning to establish a 10-year "Intelligent Manufacturing System Project," to begin in 1991, with a budget of $1 billion. It is reported by the Japanese press that both the USA and the EEC have agreed to join. Unlike previous MITI sponsored projects, it will be open from the outset to foreign partners and has the explicit objective of helping to transfer Japanese technology to others. According to Yuji Furukawa of Tokyo Metropolitan University, one of the program's architects, "Japan is keenly aware of its responsibility to contribute to the development of advanced engineering technologies and wants to share those benefits with other nations" (*New Scientist*, Feb. 3, 1990, p. 40). The deputy director of MITI's Industrial Machinery division, Kenzo Inagaki, added, "Japan can no longer continue just taking a bigger piece of world market. We risk being isolated unless we share a bit of the magic of manufacturing."

In other words, the fact that consumer products are easy to imitate is not really a problem of general importance. The competitiveness problem and the productivity problem for countries like the United States and the United Kingdom – and, more acutely for the CMEA countries – arises from the fact that Japan and the East Asian "Tigers" currently enjoy more efficient manufacturing technologies that are hard to imitate. And, by the same logic, the development problem of the Third World and the centrally planned economies amounts to the same thing. Those countries, too, are having difficulty imitating the formerly successful manufacturing technology of the USA, and even more trouble rivaling the success that evolved in Japan and its neighbors.

# Acronyms

| | |
|---|---|
| ABC | Activity-based cost accounting |
| AGV | Automated guided vehicle |
| AI | Artificial intelligence |
| AMH | Automated material(s) handling |
| AMT | Advanced manufacturing technology |
| AS/AR | Automatic storage/retrieval |
| ATE | Automatic text equipment |
| CAA | Computer aided assembly |
| CAD | Computer aided design |
| CAD/CAM | Computer aided design and manufacturing |
| CADD | Computer aided design and drafting |
| CAE | Computer aided engineering |
| CAI | Computer aided inspection |
| CAM | Computer aided manufacturing |
| CAM-I | CAM-International |
| CAP | Computer aided planning |
| CAPM | Computer aided production management |
| CAPP | Computer aided process planning |
| CAQ | Computer aided quality assurance control |
| CAR | Computer aided robotics |
| CAS | Computer aided simulation or computer aided software engineering |
| CAT | Computer aided testing |
| CEC | Commission of the European Communities |
| CIM | Computer integrated manufacturing |
| CMEA | Council for Mutual Economic Assistance |
| CNC | Computerized numerical control |
| CNMA | Communications network – manufacturing applications |
| CSFR | Czechoslovak Federal Republic |
| DBMS | Data base management system |
| DNC | Distributive numerical control or direct numerical control |
| DRAM | Dynamic RAM |

| | |
|---|---|
| EC | European Communities |
| ECE | Economic Commission for Europe |
| EDP | Electronic data processing |
| EEC | European Economic Community |
| ESPRIT | European Strategic Program for Research in Information Technology |
| FMC | Flexible manufacturing cell |
| FMS | Flexible manufacturing system |
| GT | Group technology |
| IC | Integrated circuit |
| IT | Information technology |
| JIT | Just in time |
| LAN | Local area network |
| LSI | Large-scale integration |
| MAP | Manufacturing automation protocol |
| MIS | Management information system |
| MPCS | Manufacturing planning and control system |
| MRP | Manufacturing resource planning |
| NC | Numerical control |
| OECD | Organisation for Economic Co-operation and Development |
| OPT | Optimized production technology |
| PA | Programmable automation |
| PC | Personal computer |
| PPC | Production, planning, and control |
| RAM | Random access memory |
| SQC | Statistical quality control |
| TOP | Technical and office protocol |
| TQC | Total quality control |
| VDU | Visual display unit |
| VLSI | Very large-scale integration |

# References

Abernathy, William J., 1978, *The Productivity Dilemma*, Johns Hopkins University Press, Baltimore, MD.

Abramovitz, Moses, 1989, "Notes on Postwar Productivity Growth: The Play of Potential and Realization," paper presented at Seminar on Contributions of Science and Technology to Economic Growth, OECD, Paris, June 5.

AFL-CIO, 1983, *The Future of Work*, August, AFL-CIO, Washington, DC.

Alexander, Arthur J., and Bridger M. Mitchell, 1985, "Measuring Technological Change of Heterogeneous Products," *Journal of Technological Forecasting and Social Change* 27(1–2) May.

Algren, Nelson, 1944, "Hank the Free Wheeler," in A. Botkin, *A Treasury of American Folklore*, Reprinted, Crown Publishers, New York, NY.

Altschuler, A., M. Anderson, D. Jones, D. Roos, and J. Woomack, 1984, *Future of the Automobile*, MIT Press, Cambridge MA.

*American Machinist*, 1977, "Metalworking: Yesterday and Tomorrow," [100th Anniversary Issue] McGraw-Hill, New York, NY.

*American Machinist*, 1980, "Machine Tool Technology," [Special Report 726] October, McGraw-Hill, New York, NY.

Åstebro, T., forthcoming, Vol. III.

Ayres, Robert U., 1984, *The Next Industrial Revolution: Reviving Industry Through Innovation*, Ballinger Publishing Company, Cambridge, MA.

Ayres, Robert U., 1985, "A Schumpeterian Model of Technological Change," *Journal of Technological Forecasting and Social Change* 27(4).

Ayres, Robert U., 1987, *Manufacturing and Human Labor as Information Processes*, RR-87-19, International Institute for Applied Systems Analysis, Laxenburg, Austria.

Ayres, Robert U., 1988a, "Complexity, Reliability and Design: Manufacturing Implications," *Manufacturing Review* 1(1) March:26–35.

Ayres, Robert U., 1988b, "Robotics, Employment Impact", in Dorf, ed., *Encyclopedia of Robotics*:419–436, John Wiley & Sons, New York, NY.

Ayres, Robert U., 1988c, "Barriers and Breakthroughs: An Expanding Frontiers Model of the Technology Industry Life Cycle," *Technovation*(7):87–115.

Ayres, Robert U., 1989, "Information, Computers, CIM and Productivity," paper presented at Seminar on Contributions of Science & Technology to Economic Growth, OECD, Paris, June 5.

Ayres, Robert U., and Jeffrey L. Funk, 1988, "The Role of Machine Sensing in CIM," *Robotics and Computer-Integrated Manufacturing* 5(1):55–71.

Ayres, Robert U., and Steven M. Miller, 1983, *Robotics: Applications and Social Implications*, Ballinger Publishing Company, Cambridge, MA.

Ayres, Robert U., and Ehud Zuscovitch, 1990, "Technology and Information: Chain Reactions and Sustainable Economic Growth," *Technovation*, in press.

Baily, Martin Neil, 1986, "What has Happened to Productivity Growth?" *Science* **234**, October 24:43–451.

Bessant, John, and Bill Haywood, 1988, *Islands, Archipelagoes and Continents: Progress on the Road to Computer-Integrated Manufacturing*, Elsevier, Amsterdam, Netherlands.

Bjork, Lars E., 1975, "An Experiment in Work Satisfaction," *Scientific American* **232**(3) March.

Blackman, A. Wade, 1973, "New Venture Planning: The Role of Technological Forecasting," *Journal of Technological Forecasting and Social Change* 5:25–49.

Blackman, A. Wade, E. Seligman, and G. Sogliero, 1976, "An Innovation Index Based on Factor Analysis," in *Technology Substitution Forecasting Techniques and Applications*, Elsevier, New York, NY.

Boothroyd, Geoffrey, 1980, *Design for Assembly-A*, Designer Handbook, University of Massachusetts Department of Mechanical Engineering, Amherst, MA.

Bouchut, Yves, and Patrick Besson, 1983, "How to Achieve Flexibility Through Work Reorganization," *The FMS Magazine*, April.

Bright, James R., 1958, *Automation and Management*, Harvard Business School, Boston, MA.

Bright, James R., 1966, *The Relationship of Increasing Automation and Skill Requirements*, National Commission on Employment and Economic Progress, Washington, DC.

Bursky, D., 1983, "New Process Device Structures Point to Million Transistor IC," *Electronic Design*, June 9:87–96.

Carter, Charles F., 1982, "Towards Flexible Automation," *Manufacturing Engineering*, August.

Chorafas, D.N., 1987 *Engineering Productivity Through CAD/CAM*, Computer Aided Engineering, Butterworths, London.

Cook, Nathan H., 1975, "Computer-Managed Parts Manufacture," *Scientific American* **232**(2) February:22–29.

Cross, Ralph E., 1980, "The Future of Automotive Manufacturing–Evolution or Revolution?" *Automotive Engineering*, July.

Cross, Ralph E., 1982, "Automation," in Salvendy, ed., *Handbook of Industrial Engineering*:7.5.1–7.5.4, John Wiley and Sons, New York, NY.

Dataquest, 1987, Dataquest Survey Centre Point, London, UK.

Denison, Edward F., 1962, *The Sources of Economic Growth in the United States and the Alternatives Before Us*, Supplementary Paper (13), Committee for Economic Development, New York, NY.

Denison, Edward F., 1967, *Why Growth Rates Differ*, Brookings Institution, Washington, DC.

Denison, Edward F., 1974, *Accounting for US Economic Growth, 1929-1969*, Brookings Institution, Washington, DC.

Denison, Edward F., 1979, *Accounting for Slower Growth*, Brookings Institution, Washington, DC.

Drucker, Peter E., 1990, "The Emerging Theory of Manufacturing," *Harvard Business Review*, May–June: 94–101.

Ebel, Karl-H., 1985, *The Impact of Industrial Robots on the World of Work*, GE.85-24427, United Nations, Economic Commission for Europe, Geneva, Switzerland.

Ebel, Karl-H., and Erhard Ulrich, 1987, *Social and Labor Effects of CAD/CAM*, Technical Report, International Labor Organization, Geneva, Switzerland.

ECE, 1986, *Recent Trends in Flexible Manufacturing*, Technical Report, Geneva, Switzerland.

*Electronic Design*, 1988, January 7.

Engelberger, Joseph F., 1980, *Robotics in Practice*, American Management Association, New York, NY.

Feigenbaum, A.V., 1983, *Total Quality Control*, Technical Report, [First published in 1967, republished by ASQC in 1983 (3rd ed.)].

Fleck, J., 1984, "Employment Effects of Robots," in *Human Factors in Manufacturing*, IFS, London.

Fleissner, P., 1987, "Zur Wissenschaftlich-technischen Revolution der Gegenwart," in P. Fleissner, ed., *Technologie und Arbeitswelt in Österreich: Trends bis zur Jahrtausendwend*, Vienna, Austria.

Freeman, Christopher, 1989, "The Nature of Innovation and the Evolution of the Productive System," paper presented at Seminar on Contributions of Science and Technology to Economic Growth, OECD, Paris, June 5.

Galbraith, John Kenneth, 1978, *The New Industrial State*, 3rd ed., Houghton Mifflin, Boston, MA.

Goldstein, S., 1981, *Uncertainty in Life Cycle Demand and the Preference Between Flexible and Dedicated Mass-Production Systems*, Carnegie-Mellon University, Pittsburgh, PA, December. [PhD Dissertation, MP81W00037, Mitre Corp.]

Griliches, Zvi, 1961, "Hybrid Corn: An Exploration of the Economies of Technological Change," *Econometrica* 25(4) October.

Griliches, Zvi, 1988, "Productivity Puzzles and R&D: Another Nonexplanation," *Journal of Economic Perspectives* 2(4).

Griliches, Zvi, 1989, "Productivity and Technological Change: Some Measurement Issues," paper presented at Seminar on Contributions of Science and Technology to Economic Growth, OECD, Paris, June 5.

Groover, M.P., 1983, "Fundamental Operations," *IEEE Spectrum* 20(5) May.

Gunn, Thomas G., 1982, "The Mechanization of Design and Manufacturing," *Scientific American* **247**(3) September.

Hall, Robert W., 1987, "Just-in-Time Manufacturing: Discovering the Real Thing," *Tocqueville Research Corporate Strategies*, February.

Ham, Inyong, 1971, "Group Technology," in Maynard, ed., *Manufacturing Engineering*, Handbook, 3rd ed., McGraw-Hill, New York, NY.

Harrington, Joseph, 1984, *Understanding the Manufacturing Process: Key to Successful CAD/CAM Implementation*, Marcel Dekker, Inc., New York and Basel, Switzerland.

Hartley, John, 1986, *Fighting the Recession in Manufacture*, IFS/McGraw-Hill, New York, NY.

Harvey, Robert E., 1983, "Group Technology: Revitalization by Computer Power," *Iron Age*, May 20:61–72.

Haustein, Heinz-Dieter, and Harry Maier, 1981, *The Discussion of Flexible Automation and Robotics*, WP-81-152, International Institute for Applied Systems Analysis, Laxenburg, Austria.

Haustein, Heinz-Dieter, and Harry Maier, 1985, *Flexible Automatisierung*, Akademie-Verlag, Berlin.

Hayes, Robert H., and Ramchandran Jaikumar, 1988, "Manufacturing Crisis: New Technologies Obsolete Organizations," *Harvard Business Review*, September–October:77–85.

Hounshell, D., 1984, *From the American System to Mass Production, 1800–1933*, Johns Hopkins University Press, Baltimore and London.

Howell, David R., 1985, "The Future Employment Impact of Industrial Robots," *Journal of Technological Forecasting and Social Change* **28**(4), December:297–310.

Hunt, H. Allen, and Timothy L. Hunt, 1983, *Human Resource Implications of Robotics*, Technical Report, W.E. Upjohn Institute for Employment Research, Kalamazoo, MI.

Hyer, Nancy and Urban Wemmerlöv, 1984, "Group Technology and Productivity," *Harvard Business Review*, July–August:140–149.

Jaikumar, Ramchandran, 1986, "Postindustrial Manufacturing," *Harvard Business Review* **64**(6):73.

Jaikumar, Ramchandran, 1989a, "From Filing and Fitting to Flexible Manufacturing: A Study in the Evolution of Process Control," in Ranta, Jukka P., ed, *Trends and Impacts of CIM*, WP-89-1, IIASA, Laxenburg Austria.

Jaikumar, Ramchandran, 1989b, "Japanese Flexible Manufacturing Systems," *Japan and the World Economy* **1**:113–143.

Johnston, Russell, and Paul R. Lawrence, 1988, "Beyond Vertical Integration – the Rise of the Value-Adding Partnership," *Harvard Business Review*, July–August:94–101.

Jorgenson, Dale W., 1988, "Productivity and Postwar US Economic Growth," *Journal of Economic Perspectives* **2**(4).

Kaplan, Robert S., 1989, "Management Accounting for Advanced Technological Environments," *Science* **245**, August 25:819–823.

Karatsu, H., 1984, "Quality Control - The Japanese Approach," in N. Sasaki and D. Hutchins, eds., *The Japanese Approach to Product Quality*, Pergamon, Oxford, UK.

Kaya, Y., 1986, *Economic Impacts of High Technology*, CP-86-8, International Institute for Applied Systems Analysis, Laxenburg, Austria.

Kondoleon, A.S., 1976, *Application of Technology-Economic Model of Assembly Techniques to Programmable Assembly Machine Configuration*, Masters Thesis, Massachusetts Institute of Technology, Cambridge, MA.

Krelle, Wilhelm, 1985, *Theorie des Wirtschaftlichen Wachsturms*, Springer-Verlag, Berlin.

Krelle, Wilhelm, 1989, *The Future of the World Economy*, Springer-Verlag, New York, NY.

Kume, H., 1984, "Quality Control in Japan's Industries," *The Wheel Extended* **14**(1) Toyota Motor Company.

Kuttner, B., 1983, "The Declining Middle," *Atlantic Monthly*, July:60-72.

Leontief, Wassily, and Faye Duchin, 1986, *The Future Impact of Automation on Workers*, Oxford University Press, New York, NY.

Linstone, Hal, and Devandra Sahal, eds., 1976, *Technological Substitution*, Elsevier, Amsterdam, Netherlands.

Maddison, Angus, 1987, "Growth and Slowdown in Advanced Capitalist Economies: Techniques and Quantitative Assessment," *Journal of Economic Literature* **25**: 649–698.

Mahajan, Vijay, and Yoram Wind, 1986, *Innovation Diffusion Models of New Product Acceptance*, Ballinger Publishing Company, Cambridge, MA.

Mansfield, Edwin, 1961, "Technical Change and the Rate of Imitation," *Econometrica* **29**(4) October:741–766.

Mansfield, Edwin, 1968, *Industrial Research and Technological Innovation*, W.W. Norton & Co., New York, NY.

Maynard, H.B., G.J. Stegmerten, and J.L. Schwab, 1948, *Methods-Time Measurement*, McGraw-Hill, New York, NY.

McKenney, J.L., and F.W. McFarlan, 1982, "The Information Archipelago–Maps and Bridges," *Harvard Business Review*, September/October.

Merchant, M. Eugene, 1962, "The Manufacturing Systems Concept in Production Engineering Research," *Annals of the CIRP* **10**:77–83.

Merchant, M., 1989a, *CIM - Its Evolution, Precepts, Status, and Trends*, Report, ECE/IIASA, Botevgrad, USSR, September, paper presented in the ECE/IIASA Seminar on CIM.

Merchant, M., 1989b, Private Communication.

Miller, Steven M., 1983, *Potential Impacts of Robotics on Manufacturing Costs within the Metalworking Industries*, PhD Dissertation, Technical Report, Carnegie-Mellon University, Pittsburgh, PA.

Mitrafanov, S.P., 1966, *Scientific Principles of Group Technology*, National Lending Library for Science and Technology, United Kingdom, English ed.

Mori, Shunsuke, 1987, *Social Benefits of CIM: Labor and Capital Augmentation by Industrial Robots and NC-Machine Tools in the Japanese Manufacturing Industry*, WP-87-40, International Institute for Applied Systems Analysis, Laxenburg, Austria.

Mori, Shunsuke, 1989, "Trends and Problems of CIM in Japanese Manufacturing Industries," in Jukka P. Ranta, ed., *Trends and Impacts of CIM*, WP-89-1, IIASA, Laxenburg Austria.

Nabseth, L., and G.F. Ray, eds., 1985, *Advanced Manufacturing Equipment Situation and Outlook*, COM(85), Commission of the European Communities, Brussels.

Nevins, J.L., and D.E. Whitney, 1978, "Computer-Controlled Assembly," *Scientific American* **238** February:62–74.

Nilsson, Nils J., 1984, "Artificial Intelligence, Employment and Income," *A.I. Magazine*, Summer:5–14.

Östberg, Olov, 1986, "People Factors of Robotics and Automation: European Views," in G. Tärnquist, B. Gyllström, J. Nilsson, and L. Svensson, eds., *IEEE Conference*, IEEE Computer Society Press, Washington, DC.

Opitz, Herman, 1970, *A Classification System to Describe Workpieces*, Vols. 1 and 2, Pergamon Press, London.

Opitz, Herman, and H.-P. Wiendahl, 1970, "Group Technology and Manufacturing Systems for Small and Medium Quantity Production," in *International Conference on Production Research*, Taylor and Francis Ltd, London.

Porat, Marc, 1977, *The Information Economy: Definition and Measurement*, Special Publication 77-12(i), Office of Telecommunications Policy, Washington, DC.

Ranta, Jukka, and Iouri Tchijov, 1990, "Economics and Success Factors of Flexible Manufacturing Systems: The Conventional Explanation Revisited," *International Journal of Flexible Manufacturing Systems* **2**(3):169–190.

Ranta, J., J.E. Ettlie, and R. Jaikumar, forthcoming, *Computer Integrated Manufacturing*, Volume V: *Managerial and Organizational Implications*.

Rogers, Everett M., 1962, *Diffusion of Innovations*, Free Press, New York.

Rose, Michael, 1978, *Industrial Behavior, Theoretical Developments Since Taylor*, Penguin Books, United Kingdom. [Originally published by Allen Lane, 1975.]

Salter, W.E.G., 1960, *Productivity and Technical Change*, Cambridge University Press, New York, NY.

Schurr, Sam H., 1984, "Energy Use, Technological Change and Productive Efficiency," in Hollander and Brooks, eds., *Annual Review of Energy*: **9**, Annual Reviews, Inc., Palo Alto, CA

Shurkin, Joel, 1984, *Engines of the Mind*, W.W. Norton and Co., New York, NY.

Tangerman, E.J., 1949, "Do Machine Tools Cost Too Much?" *American Machinist*, September 8.

Tani, Akira, 1989, "International Comparisons of Industria; Robot Penetration," *Journal of Technological Forecasting and Social Change* **35**(2–3):191–210, April.

Taylor, F.W., 1911, *Shop Management*, Harper and Row, New York, NY.

Tierney, B., D. Kagan, and L. Williams, eds., 1967, *The Industrial Revolution in England*, Random House, New York, NY.

Trevor, Malcolm, 1986, "Quality Control - Learning From the Japanese," *Long-Range Planning* 19(5):46–53.

Tylecote, R.F., 1976, *A History of Metallurgy*, The Metals Society, London.

US General Accounting Office, 1976, *Manufacturing Technology: A Changing Challenge to Improve Productivity*, Technical Report, Washington, DC.

US National Research Council Manufacturing Studies Board, 1984, *Computer Integration of Engineering Design and Production: A National Opportunity*, Report, Washington, DC.

US Office of Technology Assessment, 1984, *Computerized Manufacturing Automation: Employment, Education, and the Workplace*, OTA-CIT-235, Washington, DC.

Usui, Mikoto, 1988, *Managing Technological Change and Industrial Development in Japan*, Discussion Paper Series (380), Institute of Socio-Economic Planning, Japan.

von Tunzelmann, G., 1978, *Steam Power and British Industrialization to 1860*, Clarendon Press, Oxford.

Warnecke, H.J., 1989, Private Communications.

WCED, 1987, *Our Common Future*, Oxford University Press, New York, NY.

Wiener, Norbert, 1948, *Cybernetics: Control and Communications in the Animal and the Machine*, John Wiley and Sons, New York, NY.

Wildemann, H., 1988, *Wirtschaftswoche*(15):110.

Wilson, David R., 1981, *An Exploratory Study of Complexity in Axiomatic Design*, MIT Press, Cambridge, MA.

Wright, Paul K., and P.J. Englert, 1984, "Sensor Based Robotic Manipulation and Computer Vision in Flexible Manufacturing Cells," in *Winter Annual Meeting (Production Engineering Division)*:53-69, American Society of Mechanical Engineers, New York, NY.

Wyss, Hugo F., 1985, "Machines and Microelectronics", in D. Altenpohl, ed., *Informatization: The Growth of Limits*, Aluminium-Verlag, Dusseldorf.

Yen, D.W., and Paul K. Wright, 1982, *Adaptive Control in Machining – A New Approach Based on the Physical Constraints of Tool Wear Mechanisms* (Unpublished), [Submitted to Journal of Engineering for Industry].

# CIM Project Members and Alumni

Dr. Alexander K. ALABIAN
International Centre for
  Scientific and Technical
  Information
USSR
Term: Sep 1986–Dec 1988


Prof. Robert U. AYRES
Carnegie Mellon University
USA
Term: Jun  1986–Aug 1987
      May 1988–Aug 1988
      May 1989–Aug 1990


Mr. Luigi BODDA
ANSALDO Group
Energy Study Center (CESEN)
Italy
Term: Jan 1989–Jul 1990


Dr. Hans-Ulrich BRAUTZSCH
Hochschule für Ökonomie
"Bruno Leuschner"
GDR
Term: Dec 1985–Dec 1986


Mr. Gerrit Rombout DE WIT
The Netherlands
Term: YSSP 1987*

Dr. Pavel DIMITROV
Karl Marx Higher Institute
  of Economics
Bulgaria
Term: Mar 1987–Oct 1989


Prof. Shumichi FURUKAWA
Institute of Developing Economies
Japan
Term: Apr 1989–Mar 1991


Dr. Serguei U. GLAZYEV
USSR
Term: YSSP 1988*


Dr. William HAYWOOD
Brighton Politechnic
UK
Term: Jan 1990–Aug 1990


Mr. Antti JOKINEN
Technical Research Centre
  of Finland
Electrical Engineering Lab.
Finland
Term: YSSP 1986*
      Mar 1987–Apr 1987


Dr. Jens KAMMERATH
GDR
Term: YSSP 1988*

Mr. Heikki J. KEMPPI
Finland
Term: YSSP 1988*

Mr. Constantine M.A. KREUWELS
The Netherlands
Term: YSSP 1989*

Mr. Michal KUBIK
CSFR
Term: YSSP 1987*

Mr. Erkki P.A. LAUKKANEN
Finland
Term: YSSP 1987*

Prof. Milan MALY
Prague University of Economics
CSFR
Term: Jan  1986–Aug 1988
       May 1989–Jul  1989

Dr. Jari T. MIESKONEN
Finland
Term: YSSP 1988*
       Jan 1989–Mar 1989
       Dec 1989–Nov 1990

Prof. Shunsuke MORI
Dept. of Industrial Administration
Faculty of Science and Engineering
Science University of Tokyo
Japan
Term: May 1986–Apr 1987

Mr. Erkin R. NOROV
USSR
Term: YSSP 1987*

Mr. Antti PERMALA
Finland
Term: YSSP 1987*

Dr. Wolfgang J. POLT
Austria
Term: YSSP 1989*

Prof. Jukka-Pekka RANTA
Technical Research Centre
   of Finland
Laboratory of Electrical and
   Automation Engineering
Finland
Term: Sep 1987–Aug 1989

Dr. Roman SHEININ
VNIISI All Union Research Institute
   for Systems Studies
USSR
Term: Jun 1981–Jul  1981
       Oct 1986–Dec 1986
       Jan 1988–Feb 1988

Mr. Henrik Magnus SIMONS
Finland
Term: YSSP 1989*

Prof. Lars SJOESTEDT
Dept. of Transportation
   and Logistics
Chalmers University of Technology
Sweden
Term: Aug 1986–Oct 1987

Mr. Andrei R. STERLIN
USSR
Term: YSSP 1986*

Mr. Peter SZALAY
CSFR
Term: YSSP 1989*

Mr. Akira TANI
Mitsubishi Research Institute, Inc.
Japan
Term: Jul 1987–Sep 1988

Prof. Iouri TCHIJOV
Institute of Economics and
　Industrial Engineering
Siberian Branch of Academy
　of Sciences
USSR
Term: Jul 1986–Jul 1990

Mr. Karl THORDSON
Sweden
Term: YSSP 1987*

Ms. Gabriele TONDL
Institute of Political Economy
Vienna University of Economics
Austria
Term: Nov 1988–Dec 1989

Prof. Kimio UNO
Institute of Socio-Economic Planning
University of Tsukuba
Japan
Term: Mar 1989–Apr 1989

Dr. Pentti T. VUORINEN
Finland
Term: YSSP 1987*

Prof. Sten WANDEL
Business Logistics and Transport
　Systems (EKI)
Linköping Institute of Technology
Sweden
Term: Jan 1986–Jun 1988

Prof. Mitsuo YAMADA
MIE University
Japan
Term: Apr 1989–Mar 1990

Prof. Ehud ZUSCOVITCH
Department of Economics
Ben-Gurion University
　of the Negev
Israel
and
Université Louis Pasteur
Bureau d'Economie Theorique
　et Appliqué (BETA)
Strasbourg
France
Term: Aug 1989–Sep 1989

* YSSP = Participant in Young Scientists' Summer Program.

# External Contributors
# to the CIM Project

Dr. Alexandre Alabian
International Centre for
   Scientific and Technical
   Information
USSR

Dr. Ben Alders
TNO
Centre for Technology and
   Policy Studies
THE NETHERLANDS

Alexander P. Alexandrov, Ph.D.
General Director
Industry Development Institute
BULGARIA

Prof. Hans Andersin
VALMET
Corporate Head Office
FINLAND

Prof. Tamio Arai
Dept. of Precision Engineering
University of Tokyo
JAPAN

Prof. Jack Baranson
Illinois Institute of Technology
   Research Institue (IITRI)
USA

Prof. John Bessant
Centre for Business Research
Brighton Business School
Brighton Polytechnic
UK

Prof. Harry Boer
University of Twente
School of Management Studies
THE NETHERLANDS

Dr. Henk Bolk
Intervisie
THE NETHERLANDS

Prof. Yves Bouchut
Economie des Changements
   Technologiques
MSH
(Economics of Technological Changes)
FRANCE

Prof. Robert Boyer
CEPREMAP
FRANCE

Prof. Dr. Ernst Braun
Österr. Akademie d. Wissenschaften
Forschungsstelle für Technikbewertung
Technology Assessment Unit (TAU)
AUSTRIA

Dr. Hans-Ulrich Brautzsch
Hochschule f. Ökonomie
"Bruno Leuschner"
GDR

Dr. Peter Brödner
Kernforschungszentrum Karlsruhe
Projektträger Fertigungstechnik
FRG

Prof. Harvey Brooks
Harvard University
Division of Applied Sciences
Aiken Computation Lab.
USA

Dipl. Math. Petr Caslavsky CSc.
Industry Division
Economic Commission for
   Europe (ECE)
SWITZERLAND

Prof. Attila Chikan
Secretary General
International Society for
   Inventory Research (ISIR)
HUNGARY

Dott. Ing. Massimo Colombo
CSTS-National Research Center
Dip. di Elettronica
Politechnico di Milano
ITALY

Dr. G. M. De Gregorio
Faculty of Engineering
Rome University
ITALY

Dr. Rumen Dobrinsky
Institute of Social Management
BULGARIA

Prof. Giovanni Dosi
ITALY

Dr. Faye Duchin
Director
Institute of Economic Analysis
New York University
USA

Dr. Karl-H. Ebel
Industrial Labor Office
SWITZERLAND

Prof. Charles Edquist
Department of Technology and
   Social Change
University of Linköping
SWEDEN

Dr. John E. Ettlie
Director
Office of Manufacturing
   Management Research
University of Michigan
School of Business Administration
USA

Prof. James Fleck
Dept. of Business Studies
University of Edinburgh
UK

Prof. Christopher Freeman
Science Policy Research Unit
(SPRU)
University of Sussex
UK

Drs. B. G. M Fruijtier
IVA
Institute for Social Science
   Research
University of Tilburg
THE NETHERLANDS

Prof. Hans Fuchs
Director
Institute for Automation
  Technology
The Academy of Sciences of the
  German Democratic Republic
GDR

Dr. Jeffrey Funk
Westinghouse Electric
  Corporation
Research and Development Center
USA

Dr. Eoin Gahan
UNIDO
AUSTRIA

Prof. Vladimir Ganovski
Automation of Discrete
  Production Engineering
Lenin Technical University VMEI
BULGARIA

Dr. Sergei U. Glaziev
Central Economic Mathematical
  Institute (CEMI)
USSR

Prof. S. Ove Granstrand
Chalmers University of Technology
Department of Industrial Management
SWEDEN

Prof. H. D. Haustein
Hochschule f. Ökonomie
"Bruno Leuschner"
GDR

Prof. Dr. M. Hentzschel
Head of Faculty of Economics
Karl-Marx-Universität
Sektion Wirtschaftswissenschaften
GDR

Dr. Donald Hicks
Vice President
North Texas Commission
USA

Dr. Sven Ake Hörte
Institute for Management of
  Innovation and Technology
(IMIT)
SWEDEN

Dr. Raimo Hyötyläinen
Technical Research Centre of Finland
Electrical Engineering Lab.
FINLAND

Prof. Hisashi Ishitani
Institute of Interdisciplinary
  Study
University of Tokyo
JAPAN

Prof. Ramchandran Jaikumar
Harvard University
Graduate School of
  Business Administration
USA

Dr. Jens Kammerath
Ökonomisches Forschungsinstitut
  der Staatlichen Plankommission
GDR

Dr. Jan Karlsson
Industry and Technology
  Division
Economic Commission for
  Europe (ECE)
SWITZERLAND

Prof. Yoichi Kaya
Department of Electrical
  Engineering
University of Tokyo
JAPAN

Dr. Maryellen R. Kelley
Management and Public Affairs
School of Urban and Public Affairs
Carnegie Mellon University
USA

Dr. Nicole Kemeny
Bremer Institut f. Betriebstechnik
an der Universität Bremen
u. angewandte Arbeitswissenschaft
FRG

Dr. Heikki J. Kemppi
Economic Planning Center
FINLAND

Dr. Soshichi Kinoshita
Nagoya University
Faculty of Economics
JAPAN

Mrs. Anna Kochan
Editor-in-Chief
FMS-Magazine
FRANCE

Dr. Kari Koskinen
Technical Research Centre of Finland
Electrical Engineering Lab.
FINLAND

Dr. George L. Kovacs
Computer and Automation Institute
Hungarian Academy of Sciences
HUNGARY

Dr. Zdenek Kozar
Director General
VUSTE
Research Institute of Technology
and Economy in Mechanical
Engineering
CSFR

Prof. Dr. Horst Krampe
Hochschule f. Verkehrswesen
"Friedrich List"
GDR

Prof. Dr. Wilhelm Krelle
Institut f. Gesellschafts- u.
Wirtschaftswissenschaften der
Universität Bonn
FRG

Dr. Izabella Kudrycka
Research Centre of Statistics
and Economics GUS and PAN
POLAND

Prof. T. Kumpe
Philips International B. V.
Corp. O & E
THE NETHERLANDS

Dr. Juhani Lempiäinen, MScTech
Technical Research Centre of Finland
Laboratory of Engineering
Production Technology
FINLAND

Dr. Anneli Leppänen
Institute of Occupational Health
Ergonomics Section
FINLAND

Dr. Renato A. Levrero
Technological and Market Trends
Honeywell
Marketing and Planning Division
ITALY

Dr. Per Lindberg
Institute for Management of
Innovation and Technology
(IMIT)
SWEDEN

Prof. Patrick Llerena
Universite Louis Pasteur
Bureau d'Economie Theorique et
  Appliquee (BETA)
UA 1237 du CNRS
FRANCE

Prof. Dr. Dimitri Lvov
CEMI
Academy of Sciences of the USSR
USSR

Prof. Igor M. Makarov
Scientific Council for Robotics and
  Automated Manufacturing
Academy of Sciences of the USSR
Deputy Minister of Education
USSR

Dr. Tom Martin
Kernforschungszentrum Karlsruhe
FRG

Mr. Plamen Mateev
Director
CAD R&D Centre
BULGARIA

Mr. M. Eugene Merchant
Metcut Research Associates Inc.
Manufacturing Technology Division
USA

Dr. George Muskens
International Social Science
  Council
European Coordination Centre for
  Research and Documentation in
  Social Sciences
AUSTRIA

Dr. Leena Norros
Technical Research Centre
  of Finland
Electrical Engineering Lab.
FINLAND

Prof. Jim Northcott
Policy Studies Institute (PSI)
UK

Dr. Martin Ollus
Technical Research Centre
  of Finland (VTT/SAH)
Electrical Engineering Lab.
FINLAND

Feliks I. Peregoudov, Ph.D.
First Deputy Chairman, Minister
USSR State Committee
  for Public Education
USSR

Prof. Sergei Perminov
VNIIEPRANT
All-Union Research Institute for
  Economic Problems of Scientific
  and Technological Development
USSR

Mag. Wolfgang Polt
Österr. Akademie der
  Wissenschaften
Institut f. sozio-ökonomische
Entwicklungsforschung u.
Technikbewertung (ISET)
AUSTRIA

Dr. Alessandro Raimondi
Consorzio MIP
Politechnico di Milano
ITALY

Prof. V. Raju
Rochester Institute of Technology
School of Engineering Technology
USA

Dr. Vil Rakhmankulov
All-Union Institute for
  Systems Studies (VNIISI)
USSR

Dipl.Ing. Hans-Peter Roth
Gruppenleiter Producktionssysteme
Fraunhofer-Institut f.
  Produktionstechnik u.
  Automatisierung
FRG

Prof. Mitsuo Saito
Dept. of Economics
Kobe University
JAPAN

Dr. Casper T. M. Stokman
EIM
Economic Research Institute for
  Small and Medium Sized Business
THE NETHERLANDS

Dr. Lucia Tomaszewicz
Institute of Econometrics
  and Statistics
University of Lodz
POLAND

Prof. Yasuhiko Torii
Dept. of Economics
Keio University
JAPAN

Drs. Rob van Tulder
University of Amsterdam
Dept. of International Relations
THE NETHERLANDS

Dr. G. W. A. Van Dijk
Ministry of Economic Affairs
THE NETHERLANDS

Dr. Pentti Vartia
Research Institute of the
  Finnish Economy
ETLA
FINLAND

Dr. Graham Vickrey
Directorate for Science,
  Technology, and Industry
OECD
FRANCE

Prof. Dr. H. -J. Warnecke
Institutsleiter
Fraunhofer-Institut f.
  Produktionstechnik. u.
  Automatisierung
GDR

Prof. W. Welfe
Director
Institute of Econometrics
  and Statistics
University of Lodz
POLAND

Prof. Wirth
Technische Universität
Karl-Marx-Stadt
Sektion Technologie mvI
GDR

Prof. Paul K. Wright
Director
Robotics Research
New York University
Courant Institute of Mathematical
  Science
Dept. of Computer Science
USA

Dr. Geoffrey Wyatt
Technological Change
Research Centre
Heriot-Watt University
Economics Department
UK

Dr. Pekka Ylä-Anttila
The Research Institute of the
  Finnish Economy
ETLA
FINLAND

Dipl. -Ing. K. -P. Zeh
Fraunhofer-Institut f.
  Produktionstechnik u.
  Automatisierung
FRG

Prof. Dr. Gerfried Zeichen
Flexible Automation
Inst. f. Feinwerktechnik (IFWT)
Technische Universität Wien
AUSTRIA

Dr. Ehud Zuscovitch
Ben-Gurion University
  of the Negev
Dept. of Economics
ISRAEL

# Bibliography

## CIM Project Publications

Ayres, R.U., 1988, *Manufacturing and Human Labor as Information Processes*, RR-87-19, International Institute for Applied Systems Analysis, Laxenburg, Austria.

Ayres, R.U., 1988, "Complexity, Reliability, and Design: Manufacturing Implications," *Manufacturing Review* 1(1).

Ayres, R.U., 1988, "Future Trends in Factory Automation," *Manufacturing Review* 1(2).

Ayres, R.U., "Technology Forecast for CIM," *Manufacturing Review* 2(1).

Ayres, R.U., 1989, "US Competitiveness in Manufacturing," *Managerial and Decision Economics* Special Issue, Spring.

Ayres, R.U., and J. Ranta, 1988, "Factors Governing the Evolution and Diffusion of CIM," in *Proceedings of the 3rd IFAC/IFIP/IEA/IFORS Conference on Man-Machine Systems*, 14–16 June 1988, Pergamon Press, Oxford, UK.

Ayres, R.U., and E. Zuscovitch, 1990, "Technology and Information: Chain Reactions and Sustainable Economic Growth," *Technovation* 10(3):163–184.

Dimitrov, P., and S. Wandel, 1988, "Inventory Levels in Manufacturing Industries: A Cross National Comparison," in *Proceedings of the 5th International Working Seminar on Production Economics*, 22–26 February 1988, North-Holland, Amsterdam, Netherlands.

Maly, M., 1989, "Flexible Manufacturing Systems in Czechoslovakia," *Managing Automation* August.

Mieskonen, J., 1990, "FM Investments and Success of the Implementations: Case Studies from Small Industrial Economies," paper presented at Manufacturing Strategy – Theory and Practice Conference 26–27 June 1990, Proceeding of the 5th International Conference of the Operations Management Association, Warwick, UK.

Mieskonen, J., 1990, "FM Investments: Driving Forces and Implementation," in E. Eloranta, ed., *Proceedings of IFIP TC5/WG5*, 7th International Conference on Advances in Production Management Systems, 20–22 August 1990, Elsevier, Amsterdam, Netherlands.

Mori, S., 1989, "Macroeconomic Effects of Robotization in Japan," *Technological Forecasting and Social Change* **35**(2–3).

Ranta, J., 1988, "Innovations, Invariances, and Analogies: Comments on the Life Cycle Theory and the Forecasting of Future Trends In Flexible Manufacturing," in E. Razvigorova and J. Acs, eds., *Life Cycle Concept and Management Practice in Industry*, Proceedings of the Workshop held in Sofia, Bulgaria, 27–29 April 1987, WP-88-84, International Institute for Applied Systems Analysis, Laxenburg, Austria.

Ranta, J., 1988, "Impact Assessment of Automation Technology: Comments and Methodological Views," in *Proceeding of the 3rd IFAC/IFIP/IEA/IFORS Conference on Man-Machine Systems*, 14–16 June 1988, Pergamon Press, Oxford, UK.

Ranta, J., 1988, "Man-Machine Systems: Analysis, Design, and Evaluation," in *Proceeding of the 3rd IFAC/IFIP/IEA/IFORS Conference on Man-Machine Systems*, 14–16 June 1988, Pergamon Press, Oxford, UK.

Ranta, J., 1989, "The Impact of Electronics and Information Technology on the Future Trends and Applications of CIM Technologies," *Technological Forecasting and Social Change* **35**(2–3).

Ranta, J., K. Koskinen, and M. Ollus, 1988, "Flexible Production and Computer in Manufacturing: Recent Trends in Finland," *Computers in Industry*, **11**(1).

Ranta, J., and I. Tchijov, 1990, "Economics and Success Factors of Flexible Manufacturing Systems: The Conventional Explanation Revisited," *International Journal of Flexible Manufacturing Systems* **2**(3).

Ranta, J., and L. Tuominen, 1988, "Impacts of Industrial Automation: Importance of the Design Process," in *Proceedings of the 3rd IFAC/IFIP/IEA/IFORS Conference on Man-Machine Systems*, 14–16 June 1988, Pergamon Press, Oxford, UK.

Ranta, J., and S. Wandel, 1988, "Economies of Scope and Design of Flexibility in Manufacturing Logistic Systems," in *Proceedings of the 5th International Working Seminar on Production Economics*, 22–26 February 1988, North-Holland, Amsterdam, Netherlands.

Tani, A., 1989, "International Comparisons of Industrial Robot Penetration," *Technological Forecasting and Social Change* **35**(2–3).

Tchijov, I., 1985, "Input–Output Modeling," in I. Tchijov and L. Tomaszewitcz, eds., *Proceedings of the 6th IIASA Task Force Meeting*, Warsaw 16–18 December 1985, Springer-Verlag, Berlin.

Tchijov, I., 1985, "CIM Introduction: Some Socioeconomic Aspects," *Technological Forecasting and Social Change* **35**(2–3).

Tchijov, I., "Economic Structural Changes: The Problems of Forecasting," in: W. Krelle, ed., *The Future of the World Economy*, Springer-Verlag, Berlin

Tchijov, I., and E. Norov, 1989, "Forecasting Methods for CIM Technologies," in R. Grubbstrom, *et al.* eds., *Production Economics: State of the Art and Perspectives*, Elsevier, Amsterdam, Netherlands.

Tchijov, I., and R. Sheinin, 1989, "Flexible Manufacturin Systems (FMS): Current Diffusion and Main Advantages," *Technological Forecasting and Social Change* 35(2–3).

Tchijov, I., and I. Sytchova, 1985, "Technological Progress Analysis: Some Input–Output Aspects." in I. Tchijov and L. Tomaszewitcz, eds., *Proceedings of the 6th IIASA Task Force Meeting*, Warsaw 16–18 December 1985, Springer-Verlag, Berlin.

Yamada, M., and S. Kihoshita, 1989, "The Impacts of Robotization on Macro and Sectoral Economies Within a World Econometric Model," *Technological Forecasting and Social Change* 35(2–3).

Wandel, S., and R. Hellberg, 1987, *Transport Consequences of New Logistics Technologies*, RR-87-17, International Institute for Applied Systems Analysis, Laxenburg, Austria.

# CIM Conference Proceedings

Ranta, J., ed., 1989, *Trends and Impacts of Computer Integrated Manufacturing*, WP-89-1, International Institute for Applied Systems Analysis, Laxenburg, Austria. Contributions include:

Ayres, R.U., and J. Ranta, "Factors Governing the Evolution and Diffusion of CIM."

Kinoshita, S., and M. Yamada, "The Impacts of Robotization on Macro and Sectoral Economies within a World Econometric Model."

Maly, M., "Strategic, Organizational, and Social Issues of CIM: International Comparative Analysis."

Maly, M., and P. Zaruba, "Prognostic Model for Industrial Robot Penetration in Centrally Planned Economies."

Mori, S., "Trends and Problems of CIM in Japanese Manufacturing Industries from Recent Surveys in Japan."

Ollus, M., and J. Mieskonen, "Bases for Flexibility in a Small Country – Some Issues of the Finnish TES-Program."

Polt, W., "Some Considerations on Possible Macroeconomic Effects on Computer Integrated Manufacturing Automation."

Ranta, J., "The Impact of Electronics and Information Technology on the Future Trends and Applications of CIM Technologies."

Ranta, J., and I. Tchijov, "Economics and Success Factors of Flexible Manufacturing Systems."

Tani, A., "International Comparisons of Industrial Robot Penetration."

Tani, A., "Saturation Level of NC Machine-Tool Diffusion."

Tchijov, I., and R. Sheinin, "Flexible Manufacturing Systems (FMS): Current Diffusion and Main Advantages."

Tchijov, I., and A. Alabian, "Flexible Manufacturing Systems (FMS): Main Economic Features."

Karlsson, J., forthcoming, *Proceedings of ECE/IIASA Seminar on Computer Integrated Manufacturing*, Botevgrad, Bulgaria, 25–29 September 1989, ECE, Geneva, Switzerland. Contributions include:

Ayres, R.U., "Preview of the Results of the IIASA CIM Study."
Dimitrov, P., "Implementation Strategies and Logistics Aspects of CIM."
Maly, M., "Socioeconomic Issues of CIM – An International Comparison."
Mieskonen, J., "Use of FM Technology in the Small Open Economy: Case Study from Finland."
Ranta, J., and I. Tchijov, "Costs, Benefits, Users Characteristics and Successful Implementation Strategies of Flexible Manufacturing Systems: An International Comparison and Survey."
Tchijov, I., "Economic Aspects of FMS Diffusion in the World."

Haywood, B., ed., 1990, "CIM: Revolution in Progress. Technology, Organization, and People in Transition, *Proceedings of Final Conference*, 1–4 July 1990, WP-90-34, International Institute for Applied Systems Analysis, Laxenburg, Austria. Contributions include:

Ayres, R.U., "CIM Overview: Driving Forces and Implications."
Brautzsch, H.-U., "Macroeconomic Employment Effects of CIM-Application."
Haywood, B., "National Differences in the Approach to Integrated Manufacturing: A Case Study of FMS in the UK and Sweden."
Maly, M., "CIM in Centrally Planned Countries."
Mieskonen, J., "Flexible Manufacturing Systems: The Technology Behind the Success."
Mori, S., "The Diffusion of Industrial Robots and CIM in Japan."
Sheinin, R., "International Comparisons of FMS Diffusion: The USSR, the UK, and the FRG."
Ranta, J., "CIM: Flexible Technologies in Manufacturing."
Tchijov, I., "The Diffusion of Flexible Manufacturing Systems (FMS)."
Uno, K., "CIM and the Economy: Clues for Empirical Analysis."
Åstebro, T., "The International Diffusion of Computer Aided Design."

# INDEX